Heribert Insam   Andrea Rangger   ·   Microbial Communities

Springer

Berlin
Heidelberg
New York
Barcelona
Budapest
Hong Kong
London
Milan
Paris
Santa Clara
Singapore
Tokyo

H. Insam   A. Rangger   (Eds.)

Microbial Communities

Functional Versus Structural Approaches

With 83 Figures and 54 Tables

 Springer

Univ. Doz. Dr. HERIBERT INSAM
Mag. ANDREA RANGGER

Universität Innsbruck
Institut für Mikrobiologie
Technikerstraße 25
A-6020 Innsbruck

AGR
QR
III
.M379
1997

Gedruckt mit Unterstützung des Bundesministeriums für Wissenschaft
und Verkehr in Wien.

ISBN 3-540-62405-8 Springer-Verlag Berlin Heidelberg New York

Library of Congress Cataloging-in-Publication Data

Microbial communities : funtional versus structural approaches /
   Heribert Insam , Andrea Rangger, eds.
         p.   cm.
   Includes bibliographical references and index.
   ISBN 3-540-62405-8 (hardcover)
      1. Soil ,microbiology-- Reserarch--Methodology.   2. Microbial
ecology--Research--Methodology.   I. Insam, Heribert, 1957-
II. Rangger, Andrea.   1966-
QR111.M379 1997
579'. 1757--dc21   97-22416

©Springer-Verlag Berlin Heidelberg 1997
Printed in Germany

Cover design: Design & Production, Heidelberg
Cover photograph: Judith Ascher
Typesetting: Camera ready by authors
SPIN 10544412 31/3137 5 4 3 2 1 0 - Printed on acid-free paper

Preface

Research on decomposer communities of terrestrial ecosystems for a long time has focussed on microbial biomass and gross turnover parameters. Recently, more and more attempts are made to look beyond the biomass, and more specifically determine functions and populations on a smaller scale in time and space. A multitude of techniques is being improved and developed.

Garland and Mills (1991) triggered a series of publications on substrate utilization tests in the field of microbial ecology. Despite several promising results for different applications in different laboratories, many problems concerning the assay and the interpretation of results became evident. After individual discussions on the approach with colleagues from various laboratories we started to plan a workshop on the matter. The response on our first circular was extraordinary, and instead of a small workshop it became a meeting with almost 150 participants. The meeting was named *'Substrate use for characterization of microbial communities in terrestrial ecosystems'* (SUBMECO) and was held in Innsbruck, Austria, from Oct. 16-18, 1996. The very focussed scope attracted enthusiastic advocates of the approach, and also serious critics.

Some of the topics concerned improvements of current inoculation and incubation techniques, ranging from sample pre-treatment, inoculum density and incubation temperature to statistical data handling. New methods for calculating microbial diversity were proposed, as well as bootstrap methods that allow statistics with many variables on a relatively low number of replicates.

One of the main aims of the meeting was to link the functional approach with structural ones, like the determination of biochemical markers or DNA and RNA based methods. It was discussed under which circumstances substrate use tests are advisable, and for which studies other approaches offer much more information.

The selection of substrates offered on commercially available multisubstrate plates is not targeted at environmental samples. Unequivocally, the need for a set of substrates meeting the requirements of environmental analyses was recognized. In a follow-up workshop it was proposed to screen the data available in the different labs to come up with such a set of substrates. As a result of previous experience and the workshop, an EcoPlate was designed with only 31 substrates, but in three replications per plate. Prototypes of these plates are currently in the test phase in several labs. For further communication on the topic, an internet discussion forum was initiated (http:www.biolog.com).

The editors express their thanks to the members of the scientific committee of the meeting, Erland Bååth (Sweden), Ann Kennedy, Aaron Mills, Jay Garland, John Zak (USA), Lars Bakken, Shivcharn Dhillion (Norway), Colin Bell (Canada), Roland Psenner (Austria), Christoph Tebbe (Germany), Anne Winding (Denmark) and Bryan Griffiths (UK) for their initial ideas that helped to focus the meeting, their involvement in heading sessions and perform the reviewing process. The editors also thank the sponsors of the meeting for their support, among them the University of Innsbruck, the Austrian Federal Ministry of Science, the European Union (MICS and grant for Eastern European participants) and Tyrolean Airways. The help and support by colleagues and students of the Institute of Microbiology is also gratefully acknowledged.

The contents of this book range from agricultural topics to global change issues, they relate to pollutant effects and bioremediation studies, and they cover microbial community dynamics during decomposition and food fermentation processes. You will further find method comparisons (substrate use tests versus PLFA and genetic approaches) and, last but not least, critical evaluations. Papers focussing on methods improvement can be found in a Special Issue of the *Journal of Microbiological Methods*.

The Editors
Heribert Insam
Andrea Rangger

Contents

Relationship between Functional Diversity and Genetic Diversity in Complex Microbial Communities

Bryan S. Griffiths, Karl Ritz and Ronald E. Wheatley

Unit Soil Plant Dynamics, Cellular and Environmental Physiology Department, Scottish Crop Research Institute, Invergowrie, Dundee DD2 5DA, UK.

Keywords. Diversity indices, functional diversity, genetic diversity, microbial communities, resilience

1. Introduction

The operation of particular biological systems, whether ecosystem or microcosm, depends on the interplay of three general factors - environment, biological community structure (diversity), and biological activity (function). The role of diversity, particularly of micro-organisms, and the relationship between microbial diversity and function is largely unknown. The application of molecular biological techniques has shown that microbial communities, particularly in soil, are very complex (Torsvik *et al.*, 1990; Ritz *et al.*, 1997) and understanding this relationship is not straightforward. We set out in this paper to outline some of the concepts involved in the study of microbial diversity, and to present an overview of the relationship between diversity and function.

2. Definitions of diversity

Classical concepts of diversity involve species richness, evenness and composition (i.e. the numbers of different species present, the relative contribution that individuals of all species make to the total number of organisms present, and the type and relative contribution of the particular species present). These have been used to generate commonly used indices of diversity, such as the Shannon-Weaver index (H', Shannon and Weaver, 1949), where

$$\text{'diversity' } H' = - \sum p_i \cdot \log_e p_i$$
$$p_i = \text{proportion of } i\text{th species in total sample}$$

and more recently for nematode communities the maturity index, (MI, Bongers, 1990) where

$$MI = \sum v(i) \cdot p(i)$$
$$v(i) = \text{importance value of taxon } i$$

MI is based on the life-cycle strategies of various nematode taxa (de Goede, 1993) and was developed to detect ecological disturbance (Bongers *et al.*, 1995).

However, when considering microbial communities these parameters simply cannot be determined. There are no methods currently available, or likely to be available in the foreseeable future, that can determine the identity, richness and evenness of all microbial species present. Rather, microbial ecologists have to interpret the data that can be obtained and devise experimental approaches to overcome the technical shortcomings.

The non-cultivability of the majority of species present in environmental samples is now well established (Wagner *et al.*, 1993) and analysis of environmental DNA is being used to overcome this. However, the information that is available from the analysis of community DNA needs careful consideration in terms of diversity. Thus, while DNA reassociation of prokaryotic DNA has been used to estimate the richness of bacterial species (Torsvik *et al.*, 1990; Atlas *et al.*, 1994), it cannot give an absolute value for eukaryotic DNA because of repetitive sequences, although it may give a relative measure (Ritz *et al.*, 1997). In reality, however, the reassociation of both prokaryotic and eukaryotic DNA will be affected by species evenness, with the most common species contributing most strongly to that initial reassociation rate which is used for the determination. DNA cross-hybridisation gives a measure of the DNA that is common between samples (i.e. the similarity of DNA, which is a measure of relative species composition) and also a measure of the relative diversity of the samples (Lee and Fuhrman, 1990; Ritz and Griffiths, 1994). The diversity component cannot distinguish between species richness and evenness, and while it can determine the most diverse of the two samples compared it cannot give information about the absolute level of diversity involved. The interpretation of the results is also dependent on the %G+C content of the DNA (Lee and Fuhrman, 1990). A shift in the %G+C content can be used to determine changes in microbial community structure (Harris, 1994; Leser *et al.*, 1995; Griffiths *et al.*, 1997), but does not reveal any of the diversity parameters (richness, evenness, composition). The same is true of phospholipid fatty acid (PLFA) analysis, an alternative approach which also overcomes the problem of non-cultivability (Frostegård *et al.*, 1993). PCR based techniques, such as RAPD (Malik *et al.*, 1994) or denaturing gradient gel electrophoresis (DGGE, Muyzer et al., 1993), give information on species composition, and can be used to compare the common species present in samples. Unfortunately, they are not an absolute indicator of richness or the identity of the different species. Despite these problems it is likely that a combination of these techniques can reveal a great deal about microbial community diversity. An example of how changes in species richness, evenness and composition would be detected by a combination of broad-scale DNA techniques is given in Table 1.

As well as concepts of diversity related to species richness, evenness, and composition there are also approaches based on trophic diversity and functional diversity. The presence of different trophic levels is well known to be important and affect the function of the system. This was shown in experiments in which sterile media were inoculated with organisms of increasing trophic diversity, e.g.

(Woods *et al.*, 1982; Brussaard *et al.*, 1991) bacteria (B), B + bacterial-feeding protozoa (P), B + P + protozoan-feeding nematodes (N), B + P + N + nematode-feeding mites.

Table 1. Conceptual response of broad scale DNA techniques in comparing two microbial communities which differ in their species composition, evenness or richness

Technique	Communities differing in species composition	evenness	richness
Reassociation	same	same	differs
%G+C	same or differs	same or differs	differs
X-hybridization			
- similarity	<100%	<100%	<100%
- diversity	equal	unequal	unequal
PCR (i.e. RAPD, DGGE)	differs	same	differs

This has given rise to the food web approach to look at soil communities (de Ruiter *et al.*, 1993, Beare *et al.*, 1992) and simulation modelling in which changes in trophic diversity are related to ecosystem function, C and N mineralisation (de Ruiter *et al.*, 1994). Functional diversity, which is being increasingly measured with substrate utilisation assays such as Biolog (Garland and Mills, 1991), can be determined in terms of rates of utilisation, the presence or absence of utilisation, and can be determined for any number of potential substrates, as discussed elsewhere in these proceedings. The use of volatile organic compounds (VOC's) to obtain a functional profile, analogous to PLFA analysis, is another potential indicator of functional diversity (Wheatley *et al.*, 1996). It still needs to be demonstrated how these indicators of functional diversity relate to overall system function.

3. Experimental approach

The three-way interaction between :
environment - diversity - function
immediately causes problems in trying to resolve the detailed interactions between any two of the factors. Environmental conditions can affect microbial activity without affecting microbial diversity (i.e. fluctuations in warming / wetting / pH changes could directly affect process rates without altering diversity in the short-term), and many of the treatments which can be applied to alter diversity (i.e. sterilisation, addition of biocides, freezing/thawing) also affect soil environmental conditions and therefore could affect function directly. This makes it difficult to study the interactions between diversity and function, and therefore to make any predictions about the consequences of altered diversity. There are documented studies where an environmental perturbation has been shown to change diversity (e.g. Yeates and Bird, 1994), but to make the next step and predict what the effect of that altered diversity will be on ecosystem function

requires detailed experimentation of *diversity - function* interactions. To ensure that the environment component is constant means that these studies are best conducted in laboratory microcosms, from which general principles and hypotheses can be developed for testing under field conditions.

4. Examples of *diversity - function* interactions

4.1 Evenness of microbial species

Preliminary results have been obtained from a batch, liquid culture experiment (to ensure constant environmental conditions) in which a soil suspension was used to ensure as diverse a microbial inoculum as possible (Griffiths, Ritz and Wheatley, unpublished data). This diverse inocula was grown in media of different strengths, tryptone soya broth diluted 10-, 100-, and 1000- fold, sub-cultured to ensure no carry over of soil particles, and sampled at similar points in the growth curve. It was observed that although total cell numbers varied with the strength of the medium, there was no difference in species composition (as determined by denaturing gradient gel electrophoresis, Muyzer *et al.*, 1993) but an increase in species evenness (as determined from community DNA hybridisation, Ritz and Griffiths, 1994) in media of increasing strength. The function of the communities was similar, as the same spectra of volatile organic compounds, using methods described in Wheatley *et al.* (1996), were obtained and the same Biolog substrates were utilised. So, for these complex communities with the same species composition, a change in the evenness of the species had no effect on function.

4.2 Richness of microbial species

Chloroform fumigation is widely used to estimate microbial biomass, but it is known that fumigation does not result in a complete kill of all microbes. Typically, 90 - 99 % are killed (Jenkinson 1966, McGill *et al.* 1986), but values as low as 37 % have been found (Ingham and Horten 1987). Precise reasons for such variation in killing efficiency by chloroform are unclear. There appears to be no effect of matric potential or bulk density of soil upon fumigation efficiency (Toyota *et al.* 1996). Certain cells may be protected by soil organic matter or their position within the soil matrix. It was hypothesised that fumigation for varying periods of time would provide an experimental means of manipulating microbial diversity. The tenet is that fumigation for increasing periods of time will progressively destroy species that are more chloroform resistant for whatever reason. If the soils were incubated for a length of time post-fumigation, then the resulting microbial communities should be increasingly less species abundant. Aliquots of a sandy-loam soil were fumigated for up to 24h, and then incubated moist at 15°C to allow recolonisation by surviving micro-organisms. Samples were taken after 2 months to determine potential functional diversity, by the Biolog assay, and community structure, by phospholipid fatty acid analysis

(PLFA, Table 2). PLFA is not a measure of diversity but indicates differences in microbial community structure, which in this instance were most likely associated with altered species richness. Preliminary observations (Bååth, personal communication) showed that changes in PLFA pattern were not always associated with differences in Biolog substrate utilisation (Table 2). Although more definitive measures of diversity have not yet been completed, the early work does indicate that there can be changes in microbial community structure with no change in function, but that function was affected below a certain level of species richness.

Table 2. Relative changes in microbial community structure (PLFA profile), and microbial metabolic diversity (Biolog profile), in response to different times of chloroform fumigation.

Fumigation time (hrs)	PLFA profile type	Biolog profile type
0	A	1
0.5	B	1
2	C	2
24	C	3

Other approaches to reduce species richness include serial dilution, which should remove increasingly rare species, and serial filtration to remove increasingly small species. Salonius (1981) inoculated sterile forest soil (thus ensuring the same environment, a factor which compromises the fumigation approach above) with soil suspensions which had been serially diluted. After incubation for 5 months to allow recolonisation, the function of the resulting community was assessed as respiration rate when re-inoculated into sterile soil. The respiration response indicated that microbial function was not impaired until species richness was reduced below a critical level. A similar conclusion to that was drawn from the fumigation experiment.

4.3 Species composition

Long-term field plots on the same site ensure that the environment is as similar as can be achieved in the field. Fauci and Dick (1994), and Burket and Dick (1996) reported on a study of soils which have had different substrates (manure, pea residues, inorganic N, or no addition) added continually since 1931. The microbial community structures in the different plots will probably differ in species composition because of the different inputs, but they are likely to be equally diverse in terms of species richness and evenness. A Biolog assay could distinguish the pea from the manure plots (Burket and Dick, unpublished data), indicating a difference in potential metabolic diversity and possibly in species composition. The function of the communities in responding to the addition of different substrates was the same. For example, pea residues were decomposed equally well by all soils regardless of long-term history. The manure, although slower to decompose than pea residues, was also decomposed equally well by all soils regardless of long-term history. Thus, in this instance equally complex communities with different species compositions had the same function.

The experimental presence / absence of different trophic levels gave relatively easily interpretable results, as described above. It is far harder to determine the effects of diversity within a trophic level, largely because environmental or experimental effects on diversity are not confined to a single trophic level. Setala *et al.*, (1991) found that mutual relationships between fauna were important in decomposition. Vedder *et al.* (1996) defaunated field mesocosms and allowed recolonisation by microbes < 35 μm (which probably included some nematodes), mesofauna < 1mm (mainly collembola and enchytraeids) or macrofauna (also mites and earthworms). In this case final concentrations of NH_4-N and protease activity were increased by fauna but were no different between meso- and macro-fauna, whereas N mineralisation was greater in the macro than the meso-faunal treatment.

4.4 Function of individual species within a community

The concept of 'keystone' species implies that there are some species, or groups of species, that perform a singularly important ecosystem function, i.e. nitrifiers, earthworms. However, it is not simply a question of whether these species are present. The activity of these individual species is affected by the composition of the microbial community in which they reside. Thus, there was greater CO_2 evolution when barley straw was inoculated with *Phlebia radiata* alone, than with *P. radiata* together with *Trichoderma harzianum* or together with indigenous barley straw micro-organisms (Janzen *et al.*, 1995). This led the authors to conclude that interactions among microbial populations are the highest hierarchical level of control on plant residue decomposition. In a series of model experiments where chitin was degraded by two chitinolytic organisms (a fungus and a bacterium) to provide ammonium for nitrification by two nitrifying bacteria, de Boer *et al.* (1996) were able to show that for a given nitrifier population, nitrification rate was dependent on the composition of the chitinolytic community.

5. Substrate use to characterise critical levels of microbial species diversity

The few examples given above indicate that the relationships between genetic diversity and functional diversity are complex, but there is some indication that there may be a critical level of species richness below which function is impaired. We are far from being able to define and characterise what that level is and, because of the complex interactions occurring, it is unlikely that any theoretical or conceptual framework could be constructed to either simply or directly link genetic (species) diversity with specific ecosystem functions. There is, however, a greater expectation that more general indicators of ecosystem function, such as *resilience*, could be included in a theoretical framework. We define resilience as the ability of the community to withstand or recover from a perturbation. Thus, there are good reasons to expect species diversity to be directly related to resilience, but not to specific functions. Thus, if in response to a perturbation which reduces biodiversity a system maintains a range of functions, then that system can be said to be resilient. If the range of functions is impaired for a period of time, then the time taken for recovery to levels prevailing before the perturbation can be taken as a measure of resilience. Alternatively, if the system irretrievably loses some functional attributes concomitantly with the reduction in biodiversity, then it is not resilient. Substrate utilisation profiles could be used to measure a range of such functions, and the effect of reduced biodiversity on substrate utilisation used to determine resilience. This would show whether the soil community had sufficient species diversity to be able to lose some species with no measurable effect on function. Such a soil could be said to have high biological quality. If, however, function was impaired by the applied stress then its biodiversity would not be sufficient to maintain adequate biological quality. In this way substrate utilisation would be used in conjunction with applied perturbations to measure resilience and hence establish critical levels of species diversity in functional terms.

6. Acknowledgements

This work was funded by the Scottish Office Agriculture, Environment and Fisheries Department.

7. References

Atlas RM, Horowitz A, Krichevsky M, Bej AJ (1994) Response of microbial populations to environmental disturbance. Microb Ecol 22: 249-256

Beare MH, Parmelee RW, Hendrix PF, Cheng W, Coleman DC, Crossley DA Jr (1992) Microbial and faunal interactions and effects on litter nitrogen and decomposition in agroecosystems. Ecol Monogr 62: 569-591

Bongers T (1990) The maturity index: an ecological measure of environmental disturbance based on nematode species composition. Oecologia 83, 14-19

Bongers T, de Goede RGN, Korthals GW, Yeates GW (1995) Proposed changes of c-p classification for nematodes. Russ J Nematol 3: 61-62

Brussaard L, Kools JP, Bouwman LA, de Ruiter PC (1991) Population dynamics and nitrogen mineralization rates in soil as influenced by bacterial grazing nematodes and mites. In: Veeresh GK, Rajagopal D and Viraktamath CA (eds) Advances in management and conservation of soil fauna Oxford and IBH Publishing Co Pvt Ltd, New Dehli, India pp 51-523

Burket JZ and Dick RP (1996) Long-term vegetation management in relation to accumulation and mineralization of nitrogen in soils. In: Cadisch G and Giller KE (eds) Driven by nature: Plant litter quality and decomposition CAB International, Wallingford, UK pp 283-296

8

de Boer W, Klein Gunnewiek PJA, Parkinson D (1996) Variability of N mineralization and nitrification in a simple, simulated microbial forest soil community. Soil Biol Biochem 28: 203-211

de Ruiter PC, Moore JC, Zwart KB, Bouwman LA, Hassink J, Bloem J, de Vos JA Marinissen JCY, Didden WAM, Lebbink G, Brussaard L (1993) Simulation of nitrogen mineralization in the below-ground food webs of two winter wheat fields. J Appl Ecol 30: 95-106

de Ruiter PC, Neutel A-M, Moore JC (1994) Modelling food webs and nutrient cycling in agro-ecosystems. Trends Ecol Evol 9: 378-383

de Goede RGM, Bongers T, Ettema CH (1993) Graphical presentation and interpretation of nematode community structure: C-P triangles. Med Fac Landbouww Univ Gent, 58/2b: 743-750

Fauci MF, Dick RP (1994) Soil microbial dynamics: short- and long-term effects of inorganic and organic nitrogen. Soil Sci Soc Am J 58: 801-806

Frostegård Å, Tunlid A, Bååth E (1993) Phospholipid fatty acid composition, biomass and activity of microbial communities from two soil types experimentally exposed to different heavy metals. Appl Environ Microbiol 59: 3605-3617

Garland JL, Mills AL (1991) Classification and characterization of heterotrophic microbial communities on the basis of patterns of community-level sole-carbon-source utilization. Appl Environ Microbiol 57: 2351-2359

Griffiths BS, Ritz K, McNicol JW, Ebblewhite NE, Bååth E, Díaz-Raviña M (1997) The analysis and interpretation of thermal denaturation profiles from microbial community DNA, with reference to the community hybridization technique and heavy metal pollution in soil. FEMS Microbiol Ecol (submitted)

Harris D (1994) Analysis of DNA extracted from microbial communities. In: Ritz K, Dighton J and Giller KE (eds) Beyond the biomass John Wiley & Sons, Chichester, UK, pp 111-118

Ingham ER, Horton KA (1987) Bacterial, fungal and protozoan responses to chloroform fumigation in stored soil. Soil Biol Biochem 19: 54-550

Janzen RA, Dormaar JF, McGill WB (1995) A community-level concept of controls on decomposition processes: decomposition of barley straw by *Phanerchaete chrysosporium* or *Phlebia radiata* in pure or mixed culture. Soil Biol Biochem 27: 173-180

Jenkinson DS (1966) Studies on the decomposition of plant material in soil II Partial sterilization of soil and the soil biomass J Soil Sci 17: 280-302

Lee S, Fuhrman JA (1990) DNA hybridization to compare species compositions of natural bacterioplankton assemblages. Appl Environ Microbiol 56: 739-746

Leser TT, Boye M, Hendriksen NB (1995) Survival and activity of *Pseudomonas* sp Strain B13 (FR1) in a marine microcosm determined by quantitative PCR and an rRNA-targeting probe and its effect on the indigenous bacterioplankton. Appl Environ Microbiol 61: 1201-1207

Malik KA, Kain J, Pettigrew C, Ogram A (1994) Purification and molecular analysis of microbial DNA from compost. J Microbiol Meth 20: 183-196

McGill WB, Cannon KR, Robertson JA, Cook, FD (1986) Dynamics of soil microbial biomass and water-soluble organic C in Breton L after 50 years of cropping to two rotations. Can J Soil Sci 66: 1-19

Muyzer G, de Wall EC, Uitterlinden AG (1993) Profiling of complex microbial populations by denaturing gradient gel electrophoresis analysis of polymerase chain reaction-amplified genes coding for 16S rRNA. Appl Environ Microbiol 59: 695-700

Ritz K, Griffiths BS (1994) Potential application of a community hybridization technique for assessing changes in the population structure of soil microbial communities Soil Biol Biochem 26: 963-971

Ritz K, Griffiths BS, Torsvik VL, Hendriksen NB (1997) Broad-scale approaches to the determination of microbial community structure: analysis of soil and bacterioplankton community DNA by melting profiles and reassociation kinetics. FEMS Lett (in press)

Salonius PO (1981) Metabolic capabilities of forest soil microbial populations with reduced species diversity Soil Biol Biochem 13: 1-10

Shannon CE, Weaver W (1949) The mathematical theory of communication. University of Illinois Press, Urbana, Illinois, USA 117pp

Setala H, Tynismaa E, Martikainen E, Huhta V (1991) Mineralization of C, N and P in relation to decomposer community structure in coniferous forest soil. Pedobiologia 35: 285-296

Torsvik VL, Gokøyr J, Daae FL (1990) High diversity in DNA of soil bacteria. Appl Environ Microbiol 56: 782-787

Toyota K, Ritz K, Young IM (1996) Effects of soil matric potential and bulk density on the growth of *Fusarium oxysprum* f. Sp. *raphani*. Soil Biol Biochem 28: 1139-1146

Vedder B, Kampichler C, Bachmann G, Bruckner A, Kandeler E (1996) Impact of faunal complexity on microbial biomass and N turnover in field mesocosms from a spruce forest soil. Biol Fertil Soils 22: 22-30

Wagner M, Amman R, Lemmer H, Schleifer K-H (1993) Probing activated sludge with oligonucleotides specific for proteobacteria: inadequacy of culture-dependent method for describing microbial community structure. Appl Environ Microbiol 59: 1520-1525

Wheatley RE, Millar SE, Griffiths DW (1996) The production of volatile organic compounds during nitrogen transformations in soil. Pl Soil 181, 163-167

Woods LE, Cole CV, Elliott ET, Anderson RV, Coleman DC (1982) Nitrogen transformations in soil as affected by bacterial-microfaunal interactions. Soil Biol Biochem 14: 93-99

Yeates GW, Bird AF (1994) Some observations on the influence of agricultural practices on the nematode faunae of some South Australian soils. Fundam Appl Nematol 17: 133-145

Use of Molecular Probing to Assess Microbial Activities in Natural Ecosystems
Some recent examples and future perspectives

Jan C. Gottschal, Wim G. Meijer and Yasuhiro Oda

Department of Microbiology, University of Groningen, Kerklaan 30
9751 NN Haren, The Netherlands
Fax: +31.50.3632154   Email: J.C.Gottschal@Biol.RUG.NL

Our current understanding of both the natural and man-made microbial world is rudimentary. This is caused to a large extent by our inability to cultivate more than a tiny fraction of the bacteria which can be seen to live in soils, oceans and freshwater ecosystems alike. In most cases it is by no means obvious why this should be so, but it is likely that the exact reasons will be different for the various habitats. Apart from the simple fact that a significant fraction of the cells observed in the field are truly dead, more likely reasons are that microorganisms are so eminently adapted to and dependent on the physicochemical, nutritional and biological conditions in their specific environments that we are bound to fail in attempting to mimic such conditions in the laboratory. Until quite recently this situation implicated that our perception of most natural microbial communities was based entirely on those microbial species which were sufficiently flexible to adapt to the artificial conditions created in the laboratory. (Un)fortunately, we now begin to see how wrong this perception has been, as it relied on species which mostly were not representative of the dominant populations in nature. This insight was achieved by the enormous progress in molecular genetics and the subsequent advent of an almost new discipline within microbiology: **molecular microbial ecology**. The power of this novel field of microbiological research is that substantial information about the structure, the dynamics and the physiological potential can now be inferred from genetic analysis of the microbes in samples from natural habitats without the need to cultivate them (Sayler and Layton, 1990; Amann *et al.*, 1995; Pace, 1996). Using these approaches, in particular those based on the use of rRNA targeted fluorescent molecular probes, the abundance and types of microorganisms in natural communities can now be assessed and the microbial biodiversity can be surveyed comprehensively. However, it must be stressed that the results obtained in this way, do not obviate the need for measuring microbe-mediated processes.

With most molecular techniques reported to date the ecological importance of the occurrence of the myriad of different genes can **not** be assessed until their *in situ* expression can be reliably translated qualitatively and quantitatively into biochemical and biological activities. It is therefore important to note that over the past few years the application of novel molecular techniques has begun to reveal how we can obtain information about the contribution of microbial populations to natural processes without the need to cultivate them or to stimulate their *in situ* activity or growth to such an extent that we can measure their metabolic products directly. In the past a large variety of different, often highly sophisticated techniques, have been developed which presently allow the quantification of microbial processes at a macroscale level in natural systems. Using the new approaches to detecting microbial activities, predominantly based on the direct detection of metabolic gene transcripts (mRNAs) in conjunction with modern sensitive assays of metabolites and mRNA translation products (i.e. enzymes) will no doubt enormously improve our understanding of the distribution and extent of microbe-mediated processes in the field.

Below just a few selected examples will be briefly discussed of how mRNA-based detection techniques may reveal *in situ* activity of highly specifically targeted microbial populations or even individual cells.

Monitoring gene expression by direct quantification of gene transcripts

There is a strong need to predict the likely rates of biodegradation in order to successfully use *in situ* bioremediation in the cleanup of highly polluted soils.

Substantial knowledge about the pathways and degradative potentials of bacteria for many contaminating compounds is readily available. However, a major problem with obtaining reliable predictions is our current lack of knowledge about the extent of microbial biodegradative catabolic pathways in the soil environment. This is certainly true for the degradation of polycyclic aromatic hydrocarbons (PAHs), the main constituents of concern in coal tars, sludges and oils that contaminate sites such as old manufactured town-gas plants and oil refineries. In a study on gene expression in bacteria involved in the breakdown of PAHs, naphthalene degradation was taken as a model system to monitor mineralization rates along with the possible expression of a gene known to be involved in the first degradative steps (Fleming *et al.*, 1993). The naphthalene dioxygenase gene (*nahA*) is part of one of two, well characterized plasmid-borne, operons which encode the naphthalene degradation pathway in many *Pseudomonad* spp. The half life of the *nahA* mRNA transcript was reported as 12 min, long enough to permit reliable isolation from soils. The authors used an efficient total RNA extraction procedure followed by hybridization with a ^{32}P-labelled anti-sense *nahA* RNA-probe. Subsequent electrophoresis on 5% acrylamide and exposure on X-ray film allowed quantification of the *nahA* transcript by comparison with known concentrations of ^{32}P-labelled sense *nahA* standards. In this way it was shown that the soil bacterial *nahA* mRNA levels correlated strongly with the [^{14}C]naphthalene mineralization rates, suggesting that this type of analysis may also be a valid indicator of catabolic activity. Indeed, transcription analysis used in this way may be regarded as complementary to the more traditional mineralization studies. The sensitivity of the method has not been accurately determined but results from soils with less than 10^6 cfu of *nahA* containing cells per g of soil appeared unreliable. Lack of sufficient sensitivity is probably the most significant problem with the direct detection of transcripts from genes which are not very abundant. This was also illustrated in a study on gene expression, in which laboratory cultures of *Pseudomonas aeruginosa* containing the *mer*-operon encoded on the Tn501 transposon were used (Jeffrey *et al.*, 1994). Induction of this operon results in the synthesis of mercury reductase, which is responsible for detoxification of oxidized mercurial compounds by reducing and volatilizing them to metallic mercury. In this work, the sampling method used to obtain sufficient cells for analysis was studied in particular. Laboratory cultures of *Ps. aeruginosa* were collected on various filters and subjected to different lysis and extraction procedures. It was concluded that boiling the cells in an EDTA-SDS-diethyl pyrocarbonate solution allowed sufficient RNA recovery to detect *merA* transcripts by hybridization with ^{35}S-labelled anti-sense RNA probes. The method worked well with both small, large (15 cm) and cartridge-type filters allowing large quantities (up to several litres) of water to be filtered. Thus it was shown that a good correlation between extracted numbers of cells and recovery of *merA* mRNA was obtained as long as 10^6 - 10^9 cells were present. This is no problem with laboratory-grown cells in which it was shown that æmolar concentrations of oxidized mercury induced full *merA* transcription within 1 to 12 min, but for environmental samples large volumes of water will have to be filtered to obtain sufficient cells to see any expression at all. Not a very practical option for mRNA extraction from soils!

In a subsequent paper similar work was done in a mercury contaminated pond in Oak Ridge, Tennessee (Nazaret *et al.*, 1994). The most significant result of this field study was that **no** obvious direct relationship was observed between *merA* transcript concentration in the water and either the inducer of the gene (oxidized mercury) or the product of mercury reduction. Rather, the authors concluded that heterotrophic growth potential in general (based on ^{14}C-leucine incorporation) and availability of energy sources correlated best with the concentration of *merA* transcripts. Moreover, *merA* mRNA was also detected in samples in which no mercury volatilization took place at all. In the specific case of *merA* expression a possible explanation may be that induction is known to occur at nM mercury concentrations whereas the Km-values of the enzymes responsible for volatilization are in the μM-range. This is just one example of the fact that expression of genes may occur in natural samples but that numerous other factors might effectively prevent the actual enzymatic process from taking place. In other words, much like it is the case with more conventional measurements of potential activities, the detection of gene transcripts may be a necessary requirement but **not** a sufficient one.

This was also nicely illustrated in recent work on the expression of the *nprM* gene, encoding for neutral protease in *Bacillus megaterium* (Hönerlage *et al.*, 1995). Excretion of proteases has been extensively studied in laboratory cultures of soil-dwelling bacteria, but very little is known about the *in situ* expression of the genes involved. The enzymes can be detected quite well in soil samples but this provides little or no information about whether this is the result of past microbial activities or actively metabolizing and protease excreting bacteria or fungi. Transcripts of the *nprM* gene in *B. megaterium* were detected by northern blot analysis in laboratory cultures and by *in situ* hybridization of cells grown in soil microcosms. In cultures of *B. megaterium*, nprA mRNA was detected immediately from the onset of growth until the end of exponential growth. Only then a rapid increase in protease activity was observed which remained present at the same level for at least 24h after growth arrest, whereas *nprA* transcripts became nearly undetectable as soon as the cells entered the stationary phase of growth. Such a lack of direct correlation between mRNA levels and corresponding enzyme activities is not too surprising due to the simple physiological fact that mRNA is transcribed first and followed later by maximum enzyme levels detectable outside the cell. Post-transcriptional regulation may also account for the observed discrepancy between mRNA and protein levels. This was also shown for *B. megaterium* cells inoculated and grown in small moistened soil samples. In this case transcripts, which were visualized microscopically following whole-cell hybridization, were detected 4-8h prior to a rapid increase in protease enzyme activity. 24h after the onset of growth stained cells, indicative for the presence of mRNA, were no detected longer detected, whereas protease activity remained high. These results demonstrate that gene expression was detectable in single cells, using digoxigenin-based *in situ* hybridisation. However, microscopic visibility was poor, gene-expression was detected only qualitatively, and relatively high cell numbers ($> 10^6$ g^{-1} soil) had to be used.

To overcome at least the difficulties with quantification of levels of gene-transcripts a very elegant approach was introduced by Pichard and Paul (1993): **gene expression per gene dose**. Apparent (changes in) levels of mRNA are influenced by many different physiological and environmental factors other than just the degree of gene-expression itself, e.g. changing numbers of transcriptional units, changes in mRNA stability, changes in RNA polymerase pools, differences in extraction efficiency. In pure culture studies rRNA concentrations may be used as reliable "internal standards" to normalize some of these influences. But in natural samples, containing numerous species with different rRNA contents this is not possible and hence a more specific internal standard must be used. Pichard and Paul were the first to introduce a quantification of the gene itself, from which the mRNA is transcribed, as the highly specific normalizing internal standard. The method was elegantly worked out for the expression of the *xylE* gene in pure cultures of *Vibrio* sp. strain WJT-1C (pLV1013; a temperature dependent plasmid encoding for the enzyme catechol-2,3-dioxygenase). In the course of growth of *Vibrio* sp. both *xylE* mRNA and *xylE* DNA were quantified by ^{35}S-radioactivity detection on northern and Southern blots, respectively along with a recording of growth and catechol-2,3-dioxygenase activity. A short burst of mRNA synthesis was clearly shown during the early exponential growth phase followed by a rapid increase in enzyme activity several hours later. In principle this method should work quite well in natural samples, but as was the case in the previous examples the problem of sensitivity of mRNA detection has not been solved. A minimum of 10^6 cells per litre appeared a reasonable guess of the lower limit of detection, which will usually be prohibitive for successful detection of mRNAs not present at very high levels. In other words it may be preliminarily concluded that direct detection of gene transcripts may be applicable to laboratory cultures, in cells highly concentrated by filtration of aquatic samples and qualitatively in soils and microcosms. But with the currently available techniques for reliable quantitative detection in most environments, especially in soils, specific quantitative **amplification** is required to obtain meaningful information about the expression of selected genes. The most obvious approach will be the use of reverse transcriptase (RT) dependent PCR-amplification.

However, there seems to be one interesting exception to this rule: the direct detection of the RubisCO-gene transcripts, encoding for the CO_2-fixing enzyme ribulose-1,5-bisphosphate carboxylase. This enzyme is claimed to be the most abundant protein on earth and its mRNA is also present at such high levels in diatoms, algae, cyanobacteria and many autotrophically grown bacteria that direct detection constitutes no problem at all. Although the enzyme plays a minor role in soils, two elegant studies are very briefly discussed here as an example of the fascinating possibilities they provide for obtaining detailed information on the activity of individual species, their localization in the ecosystem and the phylogenetic position of members of multispecies communities.

Paul and Pichard (1996) designed ^{35}S-labelled RNA-probes, specific for the large subunit of RubisCO (*rbcL*) with sufficient sequence differences to discriminate between *Synechococcus*-type RubisCO from cyanobacteria, green algae and higher plants, and diatom-type RubisCO from chromophytic algae (including some diatoms) chrysophytes, brown and red algae and prymnesiophytes. Using these two broad-specificity probes it was shown in samples from the water column in the Gulf of Mexico that *Synechococcus*-type cyanobacterial mRNA was most abundant in the top 3m of the water, whereas at a depth of 80m the levels of diatom-like *rbcL* mRNA were highest. These abundances paralleled beautifully the observed [14]CO_2 fixation patterns and the numbers of cyanobacteria in the top layer. At the deep layer the high mRNA levels coincided with high CO_2-fixing activity but the nature of the organisms responsible for it has not been revealed yet. In a further attempt to identify the nature of the unknown organisms present at 80m depth, primers were designed based on conserved regions of the *rbcL* sequence. Using these primers in RT-PCR amplification of natural *rbcL* mRNA isolated from the deep layers, sequences with a length of 483-489 bp were obtained. These were compared with known sequences which were brought together in a phylogenetic tree of the large subunits of RubisCo. These comparisons indicate that the observed *rbcL*expression and hence CO_2-fixation at 80m depth is caused by organisms closely related to autotrophic manganese oxidizing bacteria isolated from marine environments.

This highly sophisticated use of RT-PCR amplified sequence data to identify directly the nature of organisms responsible for certain metabolic activities (detected on the basis of gene-expression!) in natural habitats is perhaps one of the most promising results of recent studies in molecular ecology.

Very recently a similar study was published (Xu and Tabita, 1996) in which ^{35}S-labelled anti-sense RNA probes were used to evaluate the contribution of diatom-type and cyanobacterium-type RubisCO to the light dependent CO_2-fixing activity in Lake Erie, one of the Great Lakes in the northern US. Quantification was based on gene expression per gene dose, using northern blot analysis. A gradual shift from diatom-type to cyanobacterium-type RubisCO activity was indicated going from shore to deeper areas of the lake. Much like in the previous study, RT-PCR was used to generate cDNA (approx. 500 bp) fragments from the various *rbcL* transcripts. Subsequently a total of 21 RT-PCR generated fragments were separately cloned and, based on their sequences, the derived amino acid sequences were used to construct a phylogenetic tree together with 13 additional sequences from known phototrophic species. This result again highlights the remarkable power of RT-PCR to determine the likely phylogenetic position of those RubisCO containing species actively involved in CO_2-fixation in the natural environment.

The use of PCR-amplification for increased sensitivity in detecting gene expression

As indicated above for the degradation of PAHs, an important incentive for studying the expression of genes comes from the need to monitor and optimize microbial degradation and remediation processes in the field.

The white rot basidiomycete *Phanerochaete chrysosporium* is extensively studied as it may be used in bioremediation of contaminated soils. The fungus is known for its ability to produce extracellular enzymes which can degrade various PAHs. This has been substantiated in several field trials. The organism possesses a complex set of peroxidases such as the lignin peroxidase (LiP) and the manganese peroxidases (MnP), each involved in degradation of different classes of PAHs. At least 10 different structurally related genes code for the LiP proteins, also several MnP-isozymes have been detected in submerged cultures and 3 *mnp*-genes have been characterized. Unfortunately, virtually all present knowledge in the expression of these genes has been obtained in chemically defined liquid cultures and almost no information is available on the pattern of enzyme expression during growth on solid substrates and in soils. Such studies are seriously hampered by difficulties encountered in isolating peroxidases from solid substrates. In a very recent paper (Bogan *et al.*, 1996) the authors described their attempts to get around this problem by studying the expression of these enzymes (i.e. manganese peroxidases) in bench-scale soil reactors by using competitive RT-PCR analysis of *mnp* transcript levels. This approach pairs the advantage of the power of PCR-amplification of mRNA with the quantitative accuracy of the competitive primer methodology (Leser, 1995). Fluorene was used in this study as a model compound which was transformed by MnP-dependent lipid peroxidation. In the soil reactor, fluorene disappeared rapidly over the first 5-15 days to very low levels, whereas the transcripts of all three mnp genes sharply peaked between days 3 and 6. Since the transcripts were PCR-amplified, detection was possible on agarose gels stained with ethidium bromide, visualized with UV light and quantified with an image analysis system. As a general measure for growth and metabolism, two more gene transcripts i.e. encoding for -tubulin (*tub*) and glyceraldehyde-3-phosphate dehydrogenase (*gpd*) were quantified as well. Both these transcripts peaked at the same time as the peroxidase transcripts. An interesting further observation was the fact that ergosterol, as a measure for fungal biomass, did **not** show such a clear correlation: as expected, a rapid increase was followed by a relatively constant high level over the remaining 20-25 days. Finally, assaying the total MnP-enzyme activity indicated a maximum several days after the peak in gene transcripts followed by a steady decline to undetectable levels around day 15. These results clearly demonstrated that fluorene disappearance was correlated with a burst of gene transcription and that enzyme activity increased roughly at the same time but remained present over a longer period. It was not shown whether a new addition of fluorene shortly after the transcripts peak resulted again in a burst of transcription or would be transformed by the enzymes still present. This work also nicely demonstrated that in the extremely heterogeneous soil environment enzyme expression could be monitored quite selectively and accurately by using PCR-based probing technology.

The above few examples of detecting gene expression by directly analysing gene transcripts (mRNA) in laboratory cultures and in environmental samples may serve to demonstrate the almost unlimited possibilities of analysing the phylogenetic composition and metabolic activities of natural communities. However, what these sophisticated procedures still fail to do, is to inform us about the activities of **individual** cells within (natural) communities. In contrast to molecular probes based on rRNA sequences which make selected individual cells visible under the (fluorescence) microscope, mRNA based probes fail to do so because of the much lower concentrations of these transcripts inside the cell. But very recent developments in this field seem to indicate that within the very near future we will be able to selectively target fluorescently labelled probes to cells with the type of mRNA transcribed from the particular gene of our choice. This possibility is suggested by the fascinating results obtained by Hodson *et al.*(1995). These authors reported a successful Prokaryotic *In situ* (PI)-RT-PCR procedure for the amplification of gene transcripts **inside** microbial cells. After sufficient amplification of a mRNA species in the presence of labelled nucleotides (digoxigenin), individual cells could be visualized microscopically. The procedure is currently further developed to make it applicable to cells collected on filters and hence make the procedure more suitable for environmental samples. First applications will no doubt be with samples from aquatic habitats, as the problems associated with applying this technology to microorganisms in the soil may still seem virtually unsurmountable.

Conclusions

Bridging the gap between microbiology in the field and in the laboratory has always been a major concern to microbial ecologists. During the past decades no doubt enormous progress has been made in our understanding of how microbes function in response to their immediate surroundings by studying them in pure and mixed cultures in the laboratory. And with the advent of new, sensitive and often highly specific techniques for detecting microbial activities in the field our understanding of how microbes probably behave in their natural environment has grown tremendously.

But it is probably the explosive growth of the number and quality of the tools of molecular genetics which now contributes most to our possibilities of studying the behaviour of microbes in their natural environment. One of these tools is the detection of mRNA levels in natural samples and in the near future possibly even in individual cells. In summary the following points may perhaps indicate why this technique is so valuable:

- if combined with PCR amplification it will often be more sensitive than detection of the actual enzyme activity itself
- it provides information about the metabolic state of the microbes at the moment of measurement rather than a view of its past accomplishments (synthesis of certain enzymes which are still there)
- it provides in principle a better reflection of the dynamics of a microbial community because of the rapid changes in mRNA levels relative to the much slower changes in cell biomass and enzyme activity
- if known engineered organisms are used the recording of gene expression provides information about the actual environmental conditions which allow particular metabolic activities or pathways to be expressed
- if combined with careful analysis of the cloned sequences of amplified gene transcripts very detailed information may be obtained about the species composition of the communities involved in a particular metabolic activity.

References

Amann RI, Ludwig W, Schleifer KH (1995) Phylogenetic identification and *in situ* detection of individual microbial cells without cultivation. Microbiol Rev 59:143-169

Bogan BW, Schoenike B, Lamar RT, Cullen D (1996) Manganese peroxidase mRNA and enzyme activity levels during bioremediation of polycyclic aromatic hydrocarbon-contaminated soil with *Phanerochaete chrysosporium*. Appl Environ Microbiol 62:2381-2386

Fleming JT, Sanseverino J, Sayler GS (1993) Quantitative relationship between naphthalene catabolic gene frequency and expression in predicting PAH degradation in soils at town gas manufacturing sites. Environ Sci Technol 27:1068-1074

Hodson RE, Dustman WA, Gang RP, Moran MA (1995) *In situ* PCR for visualization of microscale distribution of specific genes and gene products in prokaryotic communities. Appl Environ Microbiol 61:4074-4082.

Hönerlage W, Hahn D, Zeyer J (1995) Detection of mRNA of *nprM* in *Bacillus megaterium* ATCC 14581 grown in soil by whole-cell hybridization. Arch Microbiol 163:235-241

Jeffrey WH, Nazaret S, Haven RV (1994) Improved method for recovery of mRNA from aquatic samples and its application to detection of *mer* expression. Appl Environ Microbiol 60:1814-1821

Leser TD (1995) Quantitation of *Pseudomonas* sp.strain B13 (FR1) in the marine environment by competitive polymerase chain reaction. J Microbiol Meth 22:249-262

Nazaret S, Jeffrey WH, Saouter E, von Haven R, Barkey T (1994) *merA* gene expression in aquatic environments measured by mRNA production and Hg(II) volatilization. Appl Environ Microbiol 60:4059-4065

Pace NR (1996) New perspective on the natural microbial world: molecular microbial ecology. ASM-News 62:463-470

Paul JH, Pichard SL (1996) Molecular approaches to studying natural communities of autotrophs. In: Lidstrom ME, Tabita FR (eds) Microbial Growth on C1 Compounds, Kluwer Academic Publishers, Dordrecht, pp. 301-309

Pichard SL, Paul JH (1993) Gene expression per gene dose, a specific measure of gene expression in aquatic microorganisms. Appl Environ Microbiol 59:451-457

Sayler GS, Layton AC (1990) Environmental application of nucleic acid hybridization. Ann Rev Microbiol 44:625-648

Xu HH, Tabita FR (1996) Ribulose-1,5-bisphosphate carboxylase/oxygenase gene expression and diversity of Lake Erie planktonic microorganisms. Appl Environ Microbiol 62:1913-1921

Phenetic and Genetic Analyses of Bacterial Populations in Fermented Food and Environmental Samples

Henk W van Verseveld, Wilfred FM Röling, Diman van Rossum, Anniet M Laverman, Stef van Dijck, Martin Braster and Fred C Boogerd.

Department of Microbiology, Faculty of Biology, Vrije Universiteit, De Boelelaan 1087, 1081 HV Amsterdam, The Netherlands. email: verse@bio.vu.nl

Abstract. Lactic acid bacterium *Tetragenococcus halophila* is present in both Chinese and Japanese-type fermented soy sauces in Indonesia. Populations found in the Japanese-type soy sauce were more heterogeneous than found in Chinese type regarding substrate utilization. Random amplified polymorphic DNA (RAPD) revealed that heterogeneous populations at the Japanese-type industrial producer could be derived from maximally three different 'mother' strains. Genetic relatedness among isolates from different soy sauce producers was low, but protein fingerprinting indicated that all isolates still belonged to one species.

Seventeen *Bradyrhizobium* sp. strains from Zimbabwe and one *Azorhizobium* strain were compared on the basis of five genetic and phenetic features: (I) partial sequence analysis of 16 S rRNA, (ii) RAPD analysis, (iii) protein fingerprinting, and (iv) substrate utilization. The partial 16 S rDNA sequences of the seventeen *Bradyrhizobium* sp strains could be grouped in only two rDNA homology groups, and form a tight similarity cluster with *Rhodopseudomonas palustris*, *Nitrobacter* species, *Afipia* species, and *Blastobacter denitrificans* but they were less similar to other members of the *Rhizobiaceae* family. Clustering on basis of the other features revealed more diversity, as also found with *Tetragenococcus halophila* when RAPD was compared with substrate utilization.

The flux of nitrogen in an ammonia-polluted pine forest soil has been determined, and nitrate production has been shown to be due to autotrophic nitrification. Therefore, characterization on basis of substrate utilization cannot be applied. Specific primers for ammonia oxidizers to amplify 16 S rDNA have been used to establish the presence of ammonia utilizing nitrifiers in different soil horizons. Further characterization on species level by Denaturing Gradient Gel Electrophoresis showed the presence of *Nitrosospira* species.

Keywords. phenotypic, genotypic, classification, rRNA, RAPD, protein fingerprinting, GelCompar, substrate utilization, BIOLOG, soil, nitrification.

1. General Introduction

The determination and monitoring of ecological biodiversity becomes increasingly important in microbial ecological research. Specially for biorestauration of polluted sites (industrial waste water, land fills, etc.) Determination of the "in situ" biorestauration capacity is of utmost importance. Cleaning existing land fills in the netherlands (about 3000) in a traditional manner will cost the dutch community approximately 50 billion dutch guilders in the near future. Thus, also for cost reasons there is an increase in interest in the intrinsic bioremediation capacity of different polluted and unpolluted environments.

The assessment of this intrinsic bioremediation capacity of the 'in situ' microbial community can be estimated using a combination of phenetic- and genetic-oriented techniques, such as:

(i)   Determination of substrate utilization patterns e.g. by the use of the BIOLOG-system (phenetic).

(ii)   Isolation of DNA (genetic) and establishment of patterns via Random Amplified Polymorphic DNA (RAPD), and/or Restriction Fragment Length Polymorphism (RFLP).

(iii) Detection of biodiversity by PCR amplification of 16S rDNA, using general eubacterial primers, and subsequent analysis with Denaturing Gradient Gel Electrophoresis (DGGE).

(iv)   Establishment of the presence of known micro-organisms by amplifying 16S rDNA by PCR, using species-specific eubacterial primers, and subsequent analysis with e.g. Denaturing Gradient Gel Electrophoresis (DGGE)

A combination of above-mentioned methods will result in a pattern that can be specific for a certain degradation-activity/capacity, and when found back in another environment, in combination with known environmental circumstances, will possibly have a predictive value about the capacity there.

This presentation will give an overview of the successful application of the mentioned techniques in the determination of variability in food and environmental populations (section 2 and 3). Section 3.2 emphasizes the start of research in polluted areas, which at this moment is mostly directed to communities involved in N-fluxes in the environment, but in the very near future will be extended to the total microbial community that is involved in C-, N-, and S-fluxes, and degradation of xenobiotics.

2. *Tetragenococcus halophila* in Indonesian soy sauce

2.1 Introduction

The lactic acid fermentation during the brining stage in Japanese- and Chinese type soy sauce production is mainly performed by the lactic acid bacterium *Tetragenococcus halophila*, until recently known as *Pediococcus halophilus* (Int. J. Syst. Bacteriol. 44, 370-371 (1994)).

T. halophila populations in Japanese soy mash fermentation are very heterogeneous regarding carbohydrate fermentation (Uchida, 1982), metabolism of organic acids and amino acids (Kanbe & Uchida, 1982; Uchida, 1989), phage resistance (Uchida & Kanbe, 1993) and plasmid profiles (Kayahara *et al.*, 1989). About the *T. halophila* populations in Indonesian soy sauce production very little is known, while this Chinese type soy sauce manufacturing differs considerably from the production in Japan. In Indonesia, traditionally only soybeans (containing 20% sugars) are used. During growth of *T. halophila*, glucose is completely consumed and no fermentable sugars for alcoholic fermentation remain (Röling *et al.*, 1994a). The difference in raw materials used, when compared with Japanese soy sauce preparation, may have influence on *T. halophila* populations and their activities. Furthermore, large differences in traditional Indonesian soy sauce manufacturing exist, having strong implications for the brine fermentation (Röling *et al.*, 1994a-c) and possibly also for *T. halophila* populations. Seventy-two isolates (reference strains DSM20337, DSM20338, DSM20339, isolates from the traditional Chinese type soy sauce producers Libra (LIB1 to LIB10), Ikan Lele (IL1 to IL6), Purwokerto (PUR1 to PUR10) and an industrial Japanese type soy sauce producer on Java (JV1 to JV43) were characterized (Röling & Van Verseveld, 1996)

2.2 Characterization of *Tetragenococcus halophila* populations

2.2.1. Substrate utilization

All 72 isolates of *T. halophila* were homofermentative, and capable of heterolactic fermentation. Maximum specific growth rates were 0.25 ± 0.02 h^{-1}.

The method of Uchida (1982) and commercially available BIOLOG microplates were used for characterization of substrate utilization.

Inoculum preparation for the BIOLOG microplate tests were slightly adapted, since only minor aerobic growth occurred on agar plates. All 72 isolates were grown in TSB with 5% NaCl, and resuspended in a 5% NaCl solution for inoculation. This enabled comparison with the method of Uchida (1982) while growth in 5% NaCl was better than in medium without NaCl. Substrate utilization was comparable in both systems, with the exception that α-D-glucose in the microplates surprisingly seldom gave a positive reaction in the BIOLOG system, while D-fructose and D-mannose were not always positive and showed day-to-day variation. Tested according to Uchida (1982), these three substrates always showed positive reactions.

Besides α-D-glucose, D-fructose and D-mannose, all 72 isolates were also able to utilize N-acetyl glucosamine, arbutin, cellobiose, D-galactose, gentiobiose, maltose, D-mannitol, β-methyl D-glucoside, D-ribose, salicin, methyl pyruvate, pyruvic acid and glycerol. Utilization of dextrin, amylgladin, α-methyl D-glucoside, inosine, thymidine and uridine showed day-to-day variations in the BIOLOG system. The isolates were unable to utilize α-cyclodextrin, β-cyclodextrin, glycogen, inulin, mannan, L-fucose, D-galacturonic acid, D-gluconic acid, m-inositol, D-melezitose, β-methyl D-galactoside, 3-methyl glucose, D-psicose, L-rhamnose, sedoheptulosan, xylitol, D-xylose and L-malic acid, as well as the amino acids and other organic acids in the BIOLOG plates.

Substrate utilization gave rise to 13 distinct substrate utilization (SU) patterns (Röling & Van Verseveld, 1996). SU patterns are shown in Table 1.

All isolates formed ammonia from arginine, but were not able to decarboxylate lysine, histidine, phenylalanine, aspartic acid and tyrosine.

Table 1[*] Combined RAPD patterns and SU patterns for 69 *T. halophila* isolates obtained from four Indonesian soy sauce manufacturers.

Strain(s)[a]	Pattern RAPD	SU	%[b]
LIB1, 2, 3, 4, 5, 6, 7, 8, 9, 10	LIB	SU4	100
IL1, 2, 4	IL(A)	SU5	50
IL3, 6	IL(B)	SU1	33
IL5	IL(C)	SU1	17
PUR1, 2, 3, 4, 5, 6, 7, 8, 9	PUR	SU6	90
PUR10	PUR	SU7	10
JV34, 35, 36, 40, 41	JV(A)	SU3	12
JV1		SU10	2
JV6, 25, 32		SU11	7
JV39	JV(B)	SU9	2
JV5	JV(C)	SU3	2
JV3, 20		SU11	5
JV11, 18	JV(D)	SU3	5
JV12, 23		SU11	5
JV26	JV(E)	SU3	2
JV7	JV(F)	SU3	2
JV10, 33	JV(G)	SU12	5
JV14, 15	JV(H)	SU11	5
JV9, 21, 27	JV(I)	SU3	7
JV42	JV(J)	SU8	2
JV29, 30, 31, 37, 43		SU9	12
JV2, 8	JV(K)	SU8	5
JV13, 17, 19, 24		SU9	10
JV16, 28, 38		SU10	7
JV22	JV(L)	SU13	2

[a] LIB, isolate from the traditional manufacturer Libra; IL, isolate from the traditional manufacturer Ikan Lele; PUR, isolate from the traditional manufacturer Purwokerto; JV, isolate from the industrial manufacturer.

[b] % contribution of isolates with a certain combination of RAPD and SU pattern to the population at the indicated manufacturer.

[*] Table reproduced with permission of ASM from Röling and van Verseveld (1996)

DSM strain 20337 (SU pattern 1) was isolated as *Pediococcus soyae* strain d1-2 by Sakaguchi (1958). After 40 years its ability to utilize sugars had not changed. The only difference was observed for xylose. However, since all strains in the study of Sakaguchi were able to ferment xylose, while in later studies (Uchida, 1982; Röling & van Verseveld, 1996) seldom isolates with xylose fermentation ability were encountered, we assume this to be an experimental error in the study of Sakaguchi.

2.2.2. RAPD analysis

Genomic fingerprinting by Random Amplified Polymorphic DNA (RAPD) analysis allows interspecies characterization (Goodwin & Annis, 1991; Megnegneau *et al.*, 1993). This technique was applied for our isolates, using four primers (TCACGATGCA (PHR1), AGGTCACTGA (PHR2), GTATAGCAAC (PHR3) and TAGCATGATC (PHR4)). DNA fragments with sizes between 100 bp to 1500 bp were found to be reproducible when the same amplification conditions were used in different runs and for different DNA preparations from the same isolates. Relative intensities of the fragments within a pattern was found to vary in different runs, an observation also made by other workers (Goodwin & Annis, 1991). The four primers generated 76 distinct fragments, with primers PHR1, 2, 3 and 4 amplifying 28, 19, 21 and 9 different fragments respectivley. At least 29 fragments were generated for each isolate, with a minimum of four fragments per primer.

Amplification of genomic DNA with the four primers produced resp. 13, 11, 9 and 6 distinct patterns. When the patterns were combined for each isolate, 20 different RAPD combinations were observed (Röling & van Verseveld, 1996). Gels were normalized with the GelCompar software package (version 3.0; Applied Maths, Kortrijk, Belgium) and matching bands were scored. A similarity matrix was calculated by using the Jaccard coefficient (S) and was used in unweighted pair group method using arithmetic averages (UPGMA) cluster analysis (GelCompar), as described previously by Pot *et al.* (1994). Isolates from different producers showed distinct RAPD combinations (Table 1).

The genetic relationship between isolates from different producers was low (Fig.1). All isolates from the traditional producers Purwokerto as well as Libra showed identical RAPD patterns (Table 1). Three different RAPD combinations were observed for the isolates from traditional producer Ikan Lele. Isolates with combination IL(B) and IL(C) were closely related but distinct from isolates with RAPD combination IL(A). Interestingly, the DSM20337 strain, isolated from Japanese soy mash fermentation, clustered together with isolates with RAPD combination IL(B) and IL(C) at a mean correlation level of r > 0.85 (Fig.1). The *T. halophila* population at the industrial producer is genetically diverse, the 43 isolates were distributed over 13 RAPD combinations. At a mean correlation level of r > 0.85 three clusters were observed. One group comprised RAPD combinations JV(A) to JV(I), the second contained RAPD combinations JV(J) and JV(K) while isolate JV22 formed a separate branch, JV(L). Within the two large clusters differences in RAPD patterns were only slight (Fig.1).

24

Conclusively it can be stated that isolates from traditional producers clustered more strongly to each other than to the JV isolates and DSM reference strains, with the exception of the DSM 20337, IL(B) and IL(C) RAPD and SU combinations. The JV isolates were more distinct from reference strains (all from non-Indonesian origin) than from the isolates isolated from traditional producers. Table 1 shows that SU patterns of JV-isolates are very diverse, but from figure 1, at a mean correlation level of r > 0.85, three clusters can be observed (JV(A)-JV(I); JV(J) and JV(K); and

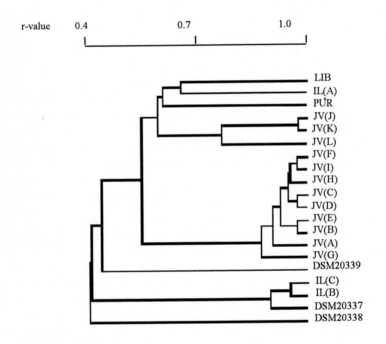

Figure 1. Dendogram showing the relationship between *T. halophila* isolates with different RAPD patterns based on UPGMA clustering of pairwise Jaccard correlation coefficients (r values) of densitograms. Figure reproduced with permission of ASM from Röling & Van Verseveld (1996).

JV(L)), indicating that the population could be derived from only three 'mother' strains. The slight genetic relationship between isolates of different manufacturers (Fig.1) raised doubts about identification as one species as was concluded from the specifics mentioned in Bergey's Manual (Schleifer, 1986). Lactic acid bacteria can easily be identified to the species level by protein fingerprinting (GelCompar). Total cell extracts of all isolates were electrophoresed and the Pearson correlation coefficients of the densitograms were calculated and clustered. All isolates grouped into one cluster at a mean correlation level of r > 0.90, indicating strongly that all isolates belong to one species of *Tetragenococcus*, *T. halophila*.

3. Prokaryotes in environmental samples

3.1 *Bradyrhizobium* strains nodulating peanut roots

The official classification of the genus *Bradyrhizobium*, as presented in Bergey's Manual of Systematic Bacteriology (Jordan, 1984), only considers phenotypic features and mole% G+C. Later, genotypic features were also described (Elkan & Bunn, 1992). Relationships between members of the *Rhizobiaceae* have been phylogenetically delineated using complete (Willems & Collins, 1993; Wong *et al.*, 1994; Yanaga & Yamasato, 1993) and partial (Young *et al.*, 1991) 16S rRNA gene (rDNA) sequences. The presentation of 16S rDNA sequence dissimilarities as phylogenetic distances is, however, based on a number of inferences and assumptions that are not without controversy. Van Rossum *et al.* (1995), therefore, present sequence dissimilarities in a purely Adansonian, numerical taxonomic fashion, treating each sequence alignment site as an operational taxonomic unit. This yields a dissimilarity dendrogram which is much less assuming than a phylogenetic tree, yet, presenting the empirical data.

Studies correlating genetic with phenetic data are essential for determining the taxonomic significance of results obtained with the various different techniques. Van Rossum *et al.* (1995) analyzed 15 *Bradyrhizobium* sp. *(Arachis)* strains capable of effectively nodulating *Arachis hypogaea* (peanut), two *B. japonicum* strains, and one *Azorhizobium caulinodans* strain from the culture collection at the Soil Productivity Research Laboratory, Marondera, Zimbabwe.

16S rDNA sequence, corresponding to *Escherichia coli* positions 44 to 337 (Brosius *et al.*, 1981), were determined for fifteen *Bradyrhizobium* strains and one *Azorhizobium* strain. The rDNA fragment for all *Bradyrhizobium* strains was 264 nucleotide long, and was 260 nucleotides long for *A. caulinodans*. All *Bradyrhizobium* strains, including two *B. japonicum* strains, gave highly homologous sequences, with differences at only 4 positions, yielding two different sequences, here designated as rDNA homology group A and B. The tree topology of Fig. 2 is consistent with rRNA:DNA hybridization studies done with members of the rRNA superfamily IV (Dreyfus *et al.*, 1988; Jarvis *et al.*, 1992), and with trees based on larger rDNA fragments (Willems & Collins, 1993; Wong *et al.*, 1994; Yanaga & Yamasato, 1993). This partial sequence thus provides enough taxonomic resolution to distinguish taxa in a highly reliable way, but a larger sequence may better resolve the finer branchings.

3.1.1. Substrate utilization

The utilization for growth of forty-three compounds by up to seventeen *Bradyrhizobium* spp. strains has been established (Van Rossum *et al.*, 1995).

Of the twenty-one aliphatic compounds tested, all strains used pentoses and hexoses for growth. Of the polyols (sugar alcohols) tested, mannitol was utilized by all strains, glycerol by almost all, while sorbitol and inositol were utilized by fewer strains. The tricarboxylic acid (TCA) cycle intermediates aconitate,

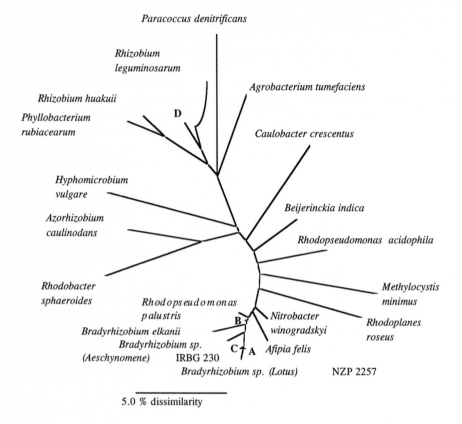

5.0 % dissimilarity

Figure 2. Dendrogram showing relationships between 23 members of the a-*Proteobacteria*, comprising all strains from the *Bradyrhizobium-Rhodopseudo-monas palustris* taxon (of which rDNA sequences were available) and other members of the family of the *Rhizobiaceae* and relatives. The tree gives a presentation of the dissimilarity matrix based on a 264 nucleotide alignment using the neighbor-joining algorithm. **A**. *Bradyrhizobium* sp. (*Arachis*) rDNA homology group A strains, *B*. sp. (*Aeschynomene*) BTAi1, *Blastobacter denitrificans* , *B. japonicum* MAR 1491, and MAR 1526; **B**. *Bradyrhizobium* sp. (*Arachis*) rDNA homology group B strains, *B. japonicum* USDA 59, *B. japonicum* LMG 6138[T], and *B. japonicum* IAM 12608[T]; **C**. *B. japonicum* RCR 3407, and *B. japonicum* USDA 110; **D**. *Rhizobium meliloti, R. fredii, Rhizobium* sp. NGR 234, and *Sinorhizobium xinjiangensis*. Figure reproduced with permission of ASM from Van Rossum *et al.* (1995)

succinate and malate were utilized by most strains (88-100%) indicating that the TCA cycle is also active in these *Bradyrhizobium* strains, as it is in other rhizobia (Stowers, 1985). Glutamate was used for growth by all strains. Di- and polysaccharides were utilized by 67% or less of the tested strains. This contrasts with the purported view that *Bradyrhizobium* strains are unable to utilize polysaccharides (Elkan & Bunn, 1992). However, Dreyfus *et al.*. (1988) already reported that some *Bradyrhizobium* strains could utilize disaccharides. The twenty-two aromatic compounds were divided into compounds that are degraded by the protocatechuate and catechol pathways. Protocatechuate was utilized as a sole carbon and energy source by all tested strains. The compounds immediately prior to protocatechuate in the degradation route, 4-hydroxybenzoate, vanillate and caffeate, were also utilized by all strains.

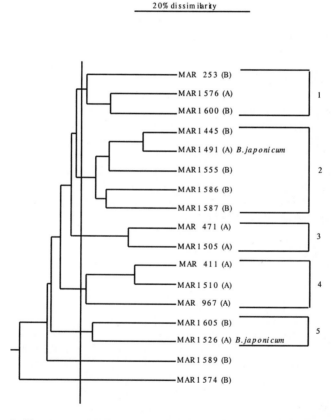

Figure 3. Phenogram showing the relationships among 17 *Bradyrhizobium sp.* strains concerning the utilization of 43 substrates, based on UPGMA cluster analysis of pairwise dissimilarities. Letters A or B in brackets denote the rRNA homology group. Figure reproduced with permission of ASM from Van Rossum *et al.* (1995)

The compounds further 'upstream' in the degradation route could not be utilized by all strains, except coniferyl alcohol and coumarate. All strains could thus utilize the key compounds in lignin degradation, namely vanillate and coniferyl alcohol, indicating metabolic adaptation to the natural habitat. 4-methyl benzoate and tyrosine could by utilized by the least number of strains. Catechol could be utilized by 12 out of 17 strains. The other compounds further 'upstream' in the catechol pathway could be utilized by fewer strains, except benzoate. Benzoate can, however, also be degraded by the protocatechuate pathway, which explains why more strains were able to grow on benzoate than on catechol. Salicylate is a product of polycyclic aromatic compound degradation. Strains able to grow on salicylate may be involved in degradation of polycyclic aromatic compounds. The degradation products of flavonoids and tannins, phloroglucinol and protocatechuic acid, can be utilized by almost all strains. However, no auxanographic growth of 5 *Bradyrhizobium* strains was observed with the flavonoids apigenin, biochanin A, chrysin, daidzein, naringenin, phloretin or quercetin. All strains could utilize adipate, an analog of b-ketoadipate, indicating the presence of an active b-ketoadipate pathway in which aromatic rings are cleaved by dioxygenases.

A numerical taxonomic analysis was performed with the substrate utilization data (Fig. 3). At or above the 84% similarity level the 17 strains can be divided into 5 clusters and 2 separate strains. Considering *Bradyrhizobium* sp. (*Arachis*) strains only, i.e. excluding *B. japonicum*, it is observed that within a cluster all strains belong to the same rDNA homology group; only MAR 1576 violates this observation. The two *B. japonicum* strains have a pairwise similarity of 83%. The cluster MAR 411/MAR 1510 (93% similar) is notable, since both strains are superior peanut microsymbionts (Van Rossum *et al.*, 1993). The cluster MAR 1586/MAR 1587 (88% similar) is interesting as both strains were recently isolated from the same site in Zimbabwe.

Four recent numerical taxonomic studies, all heavily relying on substrate utilization, had *Bradyrhizobium* strains included (Chen *et al.*, 1991; Dreyfus *et al.*, 1988; Gao *et al.*, 1994; Ladha & So, 1994). All found a clear separation of the genus *Bradyrhizobium* from other genera in the family *Rhizobiaceae*. *B. japonicum* strains always clustered together with other *Bradyrhizobium* spp. strains, except in the analysis of Dreyfus *et al.*. (1988). All reports, including this one (Fig. 3), grouped all non-phototrophic *Bradyrhizobium* strains at or above the 70% similarity level. Among the non-phototrophic *Bradyrhizobium* strains 3-5 subphena have been observed above 80% similarity (Gao *et al.*, 1994; Ladha & So, 1994): these subphena could constitute subspecies from *B. japonicum* according to the common finding that within species 80% or higher similarity occurs (Sneath, 1984).

3.1.2. RAPD analysis

Nine *Bradyrhizobium* sp. (*Arachis*) strains and two *B. japonicum* strains were fingerprinted using RAPD with 3 primers (primer 1=5'-GACGACGACGACGAC-3' (Harrison *et al.*, 1992), primer 2=5'-

TGCGCCGAATTATGCGG-3', primer 3=5'-ACGGAGTTGGAGGTC-3' (primer 2 and 3 from the primer library of the Vrije Universiteit, selected on a mole% G+C of app. 60 %)). All amplification products had lengths between 150-2250 base pairs. The primers produced up to 10 products per strain, while the number of markers across all eleven strains were 9, 20, and 13 for primers 1, 2, and 3, respectively. Primer 2 produced the most RAPD markers making it the most successful oligonucleotide in generating polymorphic genomic-DNA patterns. The banding patterns of eleven *Bradyrhizobium* strains generated with the primers 1, 2 and 3 (with a total of 42 markers) were combined per strain in one synthetic (normalized) gel as shown in Fig. 4. The densitograms of those combined patterns were pairwise analyzed by correlation, resulting in correlation coefficients (r-values) ranging from r=0.698 downward to non-significant levels below r=0.078, evidencing considerable genetic diversity. Comparison of the two marker lanes (r=0.948; Table 3), originating from two different gels, provides an estimate of the systematic error when using r-values for pairwise comparisons. Clustering analysis (Fig. 4) shows that there is only one reasonable tight cluster formed at r>0.50, consisting of 2 *B. japonicum* strains (MAR 1491 and MAR 1526) and 3 *Bradyrhizobium* sp. (*Arachis*) strains (MAR 411, MAR 471 and MAR 967). All of these strains belong to rDNA homology group A; the 6th strain belonging to this rDNA homology group (MAR 1576) is, however, the most

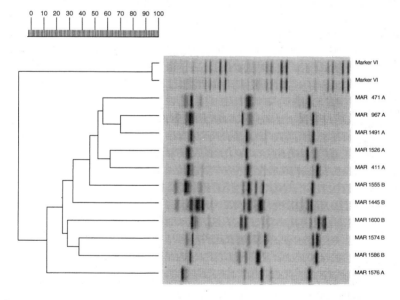

Figure 4. RAPD patterns for nine Bradyrhizobium sp. (Arachis) strains and two B. japonicum (1491 and 1526) strains produced with primers 1,2 and 3 . The gels are linked head to tail. The UPGMA clustering and r-values are shown. Letters A or B in brackets denote the rRNA homology group. Figure reproduced with permission of ASM from Van Rossum *et al.* (1995)

distantly related strain within this group of 11 strains. The strains belonging to rDNA homology group B are more polymorphic. They could only be clustered at r<0.50 (Fig. 4). The genetic diversity as revealed by RAPD fingerprinting is clearly greater than that obtained with rDNA sequence comparison. It appears again that rDNA homology is a prerequisite for RAPD profile similarity (as for the 5 clustered rDNA homology group A strains), but not a guarantee (as for the rDNA homology group B strains). RAPD profile similarity together with high rDNA sequence similarity are in our opinion good evidence for genetic relatedness of strains. Polymorphism increases with increasing the numbers of primers used, because more genomic sites are covered. The use of more primers (yielding more markers) results in an (overall) reduced correlation between profiles: profile correlations based on primer 1 only resulted in r-values between 0.80 and 1.00, while analyzing for the combined primers 1+2+3 resulted in r-values below 0.70. A large number of RAPD markers (>150) together with a sufficient number of products per strain (>10) are necessary to draw reliable (taxonomic) conclusions (Van Coppenolle et al., 1993). Such criteria were only partly met in this and in another study involving Bradyrhizobium strains (Dooley et al., 1993).

It can be stated that rDNA-based relationships only cannot be conclusive at the species level. In several, but not all, cases the rDNA clustering corresponded with clusterings based on RAPD, protein finger-printing (Van Rossum et al., 1995) or substrate utilization: it appeared that high rDNA homology (identity) is a prerequisite, but not a guarantee, for high similarity values obtained with other techniques. The rDNA homology groups A and B, as described here, were generally separately clustered in the RAPD, protein finger-printing and substrate utilization analyses, i.e., A and B strains, generally, did not occur in one cluster. In general, unacceptable inconsistencies were found across the various methods, preventing us to draw taxonomic conclusions from the combination of the four mentioned studies. The rDNA nucleotide similarities clearly show that Bradyrhizobium strains, including the phototrophic strains, cluster together with Rhodopseudomonas palustris, Nitrobacter species, Afipia species and Blastobacter denitrificans (Fig. 2).

3.2 Nitrification in sludge and dutch coniferous forest soils

The last few decades high atmospheric nitrogen input in nature, soil fertilization strategies, and increased house-hold and industrial ammonia production has caused increased nitrogen levels in waste water treatment plants, and nitrate leaching in e.g. forest ecosystems. Field experiments along a N-deposition gradient in Europe and laboratory experiments using pine forest soil cores have revealed that nitrate mobilization follows a seasonal pattern with highest values during late autumn and early winter. Ammonium availability, heterotrophic decomposition and dynamics of trophic interactions, all seem to influence the autotrophic nitrate production pattern (nitrification) in these forest soils. The importance of the stratification of the organic horizons for these processes have been studied in a coniferous forest. Concentrations of exchangeable nitrate and ammonium, pH and moisture were examined in the different layers. Nitrifying bacteria in these layers were enumerated by the most probable number technique (MPN). The highest concentrations of nitrate and ammonium were found in the litter layer, as the highest numbers of nitrifiers, whereas hardly none were present in deeper layers. Besides MPN counts, molecular techniques are used to establish the presence of nitrifiers (Laverman et al., 1996). Since nitrifiers are aerobic autotrophic organisms, only depending on the conversion of ammonia in nitrite or nitrite in nitrate for their energy supply, they are impossible to characterize by the pattern of substrate utilization except for carbondioxide utilization and N-compounds conversions.

3.2.1 Characterization of ammonia utilizing nitrifiers in municipal waste water sludge

Sludge from a nitrifying municipal waste water oxidation ditch (TNO, Delft, The Netherlands) has been studied extensively (Muller, 1994). Nitrification capacities are always very high when sludge is subcultured in a retention fermentor, due to 100% retention of biomass, and thus preventing wash-out of slow growing nitrifying organisms. Retention systems for waste water treatment are concluded to be very efficient for carbon and ammonia removal. For analysis of the nitrifying population, DNA is isolated using the standard protocol of Smalla et al. (1993) including 3 times Wizard DNA cleanup (Promega, Madisson, WI, USA). Two common eubacterial primer sets for 16S rDNA, i.e pA/pHr and pD/pEr (Edwards et al., 1989), and two specific ammonia-oxidizing bacterial primer sets, i.e. pAAOf/pAAOr (Hiorns et al., 1995) and pCTOf/pCTOr (G. Kowalchuk, NIOO, CTO, Heteren, The Netherlands), were used to amplify 16S rDNA sequences from the purified sludge DNA. All primer sets yielded products after the polymerase chain reaction with the expected number of base-pairs, strongly indicating that eubacteria are present as expected, and that indeed ammonia-oxidizing bacteria are present in sufficient amount for the direct amplification from isolated DNA. This is in contrast to DNA isolated from environmental samples in the UK (and see section 2.2.2 below), where nested PCR amplification was necessary for the detection of ammonia-oxidizing bacteria (Hiorns et al., 1995). In the near future the sludge population will be further characterized using denaturing gradient gel electrophoreses of PCR-ed DNA, using specific 16S rDNA primers with a GC-clamp (Muyzer et al., 1993).

3.2.2 Characterization of ammonia utilizing nitrifiers by DGGE

Bulk DNA was isolated from different soil horizons sampled at an ammonium-polluted coniferous forest in Wekerom, and at a less-polluted coniferous forest in Hulshorsterzand, The Netherlands, and nested PCR (Hiorns et al., 1995) with common eubacterial primers for 16S rDNA, i.e pA/pHr and pD/pEr (Edwards et al., 1989), and specific ammonia-oxidizing bacterial primers, i.e. pAAOf/pAAOr (Hiorns *et al.*, 1995) and pCTOf/pCTOr (G. Kowalchuk, NIOO, CTO, Heteren, The Netherlands) was performed to establish the presence of ammonia-oxidizers. By this technique we could prove beyond doubt the presence of autotrophic ammonium oxidizers in the litter, fragmentation, and humic layers at both sampling sites. In mineral horizon no ammonia-oxidizers were detected yet. Further identification has been done using Denaturing Gradient Gel

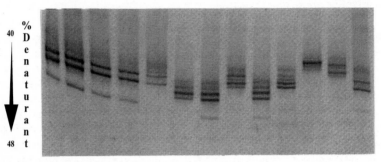

Figure 5. DGGE-patterns of DNA samples from the environment and cloned DNA, after nested PCR with common 16S rDNA primers and GC-clamped ammonia-oxidizing b-proteobacterial specific CTO primers. **Lane 1**: A fragmentation layer of Wekerom; **Lane 2**: Humic layer of Wekerom; **Lane 3**: Another fragmentation layer of Wekerom; **Lane 4**: Another fragmentation layer of Wekerom; **Lane 5**: Litter layer of Hulshorsterzand; **Lane 6**: Clone of cluster 1 Marine *Nitrosospira*; **Lane 7**: Clone of cluster 2 acid-tolerant *Nitrosospira*; **Lane 8**: Clone of cluster 3 type-strain *Nitrosospira*; **Lane 9**: Clone of cluster 4 (from enrichment cultures) *Nitrosospira*; **Lane 10**: Clone of cluster 5 (from enrichment cultures)*Nitrosomonas*; **Lane 11**: Clone of cluster 6 Marine *Nitrosomonas*; **Lane 12**: Clone of cluster 6a (from sediment)*Nitrosomonas*; **Lane 13**: Control clone of cluster 3 *Nitrosospira*.

Electrophoresis (DGGE), using the CTO-primers of the NIOO, CTO (Heteren, The Netherlands). The first results of such an exercise are shown in figure 5. When comparing lanes 1-5 (fragmentation, humus, and litter) with lanes 8 and 13 in figure 5, it will be clear that like with the UK-environmental samples (Hiorns *et al.*, 1995) we were only able to detect *Nitrosospira species* of Cluster 3 (Stephen *et al.*, 1996) of the ammonia-oxidizing bacteria in our samples.

This does not mean that no *Nitrosomonas* species are present. They presumably are present in very low concentrations beside the found *Nitrosospira* types, so that even nested PCR is not able to detect them. Enrichment cultures of the UK-environmental samples unequivocally showed the presence of *Nitrosomonas* species. Future research will be directed to further characterization to the organismal level within cluster 3 by the use of specific probes and sequencing of the PCR fragments, and the detection of other clusters, including the *Nitrosomonas* ones, by enrichment cultures.

Final remarks. Despite the differences in available substrates for *Tetrageno-coccus halophilus* in the production of traditional Indonesian soy sauce, the differences in the ability to utilize substrates were small, indicating that *T. halophila* cells are not energy-limited or introduction of new strains with better substrate utilization characteristics seldom occurs. The latter seems favored, since even at the Industrial producer with the most heterogeneous population, it is derived from only three strains through mutation and selection. For *Bradyrhizobium* strains substrate utilization patterns gave a clear separation of the genus from other genera. Although substrate utilization seems to be a good discriminative method for identification to the species level it cannot be used for autotrophic organisms like nitrifiers. Moreover, we are certain that more characteristics should be used for a final characterization on the species level. The minimum requirements should be (i) rRNA/DNA analysis, (ii) substrate utilization (or establishment of strict autotrophic characteristics), and (iii) protein profiling. In addition it is recommended to use additional methods such as RAPD and RFLP (Restriction Fragment Length Polymorphism) profiling.

In our opinion detection and characterization of micro-organisms in the environment is of extreme importance for a quantitative description of the processes in which they are involved. N-fluxes in the environment e.g. of course can be described in descriptive flux models, but such models cannot quantify and predict the influence of different biotic and abiotic variables on the conversions and their fluxes without knowledge of the organisms involved. Future work will emphasize on the strength of quantitative modeling in combination with knowledge about the population structure.

4. References

Brosius J, Dull TJ, Sleeter DD, Noller HF (1981) Gene organisation and primary structure of a ribosomal RNA operon from *Escherichia coli*. J Mol Biol 148:107-127.

Chen WX, Li GS, Qi YL, Wang ET, Yuan HL, Li JL (1991) *Rhizobium huakuii* sp. nov. isolated from the root nodules of *Astragalus sinicus*. Int J Syst Bacteriol 41:275-280.

Dooley JJ, Harrison SP, Mytton LR, Dye M, Cresswell A, Skot L, Beeching JR (1993) Phylogenetic grouping and identification of *Rhizobium* isolates on the basis of random amplified polymorphic DNA profiles. Can J Microbiol 39:665-673.

Dreyfus B, Garcia JL, Gillis M (1988) Characterisation of *Azorhizobium caulinodans* gen. nov., sp. nov., a stem-nodulating nitrogen-fixing bacterium isolated from *Sesbania rostrata*. Int J Syst Bacteriol 38:89-98.

Edwards U, Rogall T, Blocker H, Emde M, Böttger EC (1989) Isolation and direct complete nucleotide determination of entire genes: characterization of a gene coding for 16S ribosomal RNA. Nucleic Acids Res 17, 7843-7853.

Elkan GH, Bunn CR(1992) The rhizobia, p. 2197-2213. *In* A. Balows *et al.*. (ed.), The prokaryotes: a handbook on the biology of bacteria: ecophysiology, isolation, identification, applications. Springer Verlag, New York, U.S.A.

Gao JL, Sun JG, Li Y, Wang ET, Chen WX (1994) Numerical taxonomy and DNA relatedness of tropical rhizobia isolated from Hainan province, China. Int J Syst Bacteriol 44:151-158.

Goodwin PH, Annis SL (1991) Rapid identification of genetic variation and pathotype of *Leptosphaeria maculans* by random amplified polymorphic DNA assay. Appl Environ Microbiol 57:2482-2486.

Harrison SP, Mytton LR, Skot L, Dye M, Cresswell A (1992) Characterisation of *Rhizobium* isolates by amplification of DNA polymorphisms using random primers. Can J Microbiol 38:1009-1015.

Hiorns WD, Hastings RC, Head IM, McCarthy AJ, Saunders JR, Pickup RW, Hall GH (1995) Amplification of 16S rRNA genes of autotrophic ammonia-oxidizing bacteria demonstrates the ubiquity of *Nitrosospiras* in the environment. Microbiol 141, 2793-2800.

Jarvis BDW, Gillis M, De Ley J (1986) Intra- and intergeneric similarities between the ribosomal acid cistrons of *Rhizobium* and *Bradyrhizobium* species and some related bacteria. Int J Syst Bacteriol 36:129-138.

Jordan DC (1984) Family III *Rhizobiaceae* Conn 1938, p. 234-254 *In* N. R. Krieg and J. G. Holt (ed.), Bergey's manual of systematic bacteriology. Williams and Wilkins, Baltimore, U.S.A.

Kanbe C, Uchida K (1982) Diversity in the metabolism of organic acids by *Pediococcus halophilus*. Agri. Biol Chem 46:2357-2359.

Kayahara H, Yasuhira H, Sekiguchi J (1989) Isolation and classification of *Pediococcus halophilus* plasmids. Agric Biol Chem 53:3039-3041.

Ladha JK, So RB (1994. Numerical taxonomy of photosynthetic rhizobia nodulating *Aeschynomene* species. Int J Syst Bacteriol 44:62-73.

Laverman AN, Sita Murti N, van Dijck S, Zoomer HR, Braster M, Verhoef HA, van Verseveld HW (1996) Analysis of nitrifying populations in coniferous forest soils. Abstracts of the SUBMECO Conference, Innsbruck, Austria, Oct. 16-18, 1996

Megnegneau B, F Debets F, Hoekstra RF(1993) Genetic variability and relatedness in the complex group of black Aspergilli based on random amplification of polymorphic DNA. Curr Genet 23:323 - 329.

Muller EB (1994) Bacterial energetics in aerobic wastewater treatment. Ph.D. thesis, Vrije Universiteit, Amsterdam, The Netherlands.

Muyzer G, de Waal EC, Uitterlinden AG (1993) Profiling of complex microbial populations by denaturing gradient gel electrophoresis analysis of polymerase chain reaction-amplified genes coding for 16S rRNA. Appl Environ Microbiol 59, 695-700.

Pot B, Vandamme P, Kersters K (1994) Analysis of electrophoretic whole-organism protein fingerprints, p. 493-521. *In* M. Goodfellow and A.G. O'Donell (ed.), Chemical methods in prokaryotic systematics. John Wiley and Sons Ltd., Chicester, United Kingdom.

Röling WFM, Timotius KH, Prasetyo AB, Stouthamer AH, van Verseveld HW (1994a) Changes in microflora and biochemical composition during the baceman stage of traditional Indonesian *kecap* (soy sauce) production. J Ferment Bioeng 77:1-9.

Röling WFM, , Stouthamer AH, van Verseveld HW (1994b) Physical factors influencing microbial interactions and biochemical changes during the baceman stage of Indonesian *kecap* (soy sauce) production. J Ferment Bioeng 77:293-300.

Röling WFM, Schuurmans FP, Timotius KH, Stouthamer AH, van Verseveld HW (1994c) Influence of prebrining treatments on microbial and biochemical changes during the baceman stage in Indonesian *kecap* (soy sauce) production. J Ferment Bioeng 77:400 - 406.

Röling WFM, van Verseveld HW (1996) Characterization of *Tetragenococcus halophila* populations in Indonesian soy mash (kecap) fermentation. Appl Environ Microbiol 62: 1203-1207

Sakaguchi K (1958) Studies on the activities of bacteria in soy sauce brewing. Part III. Taxonomic studies on *Pediococcus soyae* nov. sp., the soy sauce lactic acid bacteria. Bull Agr Chem Soc Japan 22:353-362.

Schleifer KH (1986) Gram positive cocci, p. 899. *In* P.H.A. Sneath (ed.), Bergey's manual of systematic bacteriology, 9th ed. Williams and Wilkes Co., Baltimore.

Smalla K, Cresswell N, LC Mendonca-Hagler LC, Wolters A, JD van Elsas JD (1993) Rapid DNA extraction protocal from soil for polymerase chain reaction-mediated amplification. J Appl Bacteriol 74: 78-85

Sneath PHA (1984) Numerical taxonomy, p. 5-7. *In* N. R. Krieg and J. G. Holt (ed.), Bergey's manual of systematic bacteriology. Williams and Wilkins, Baltimore, U.S.A.

Stephen JR, McCaig AE, Smith Z, Prosser JI, Embley TM (1996) Molecular diversity of soil and marine 16S rDNA sequences related to the b-subgroup ammonia oxidising bacteria. Accepted to Appl Environ Biol

Stowers MD (1985) Carbon metabolism in *Rhizobium* species. Ann. Rev. Microbiol. 39:89-108.

Uchida K (1982) Multiplicity in soy pediococci carbohydrate fermentation and its application for analysis of their flora. J Gen Appl Microbiol 28:215-225.

Uchida K (1989) Trends in preparation and uses of fermented and acid-hydrolyzed soy sauce, p.78. *In* Proceedings of the world congress on vegetable protein utilization in human foods and animal feedstuffs, Singapore.

Uchida K, Kanbe C (1993) Occurrence of bacteriophages lytic for *Pediococcus halophilus*, a halophilic lactic-acid bacterium, in soy sauce fermentation. J Gen Appl Microbiol 39:429-437.

Van Coppenolle B, Watanabe I, Van Hove C, Second G, Huang N, McCouch SR (1993) Genetic diversity and phylogeny analysis of *Azolla* based on DNA amplification by arbitrary primers. Genome 36:686-693.

Van Rossum D, Muyotcha A, van Verseveld HW, Stouthamer AH, Boogerd FC (1993) Effects of *Bradyrhizobium* strain and host genotype, nodule dry weight and leaf area on groundnut (*Arachis hypogaea* L. ssp. *fastigiata*) yield. Plant and Soil 154:279-288.

Van Rossum D, Schuurmans FP, Gillis M, Muyotcha A, van Verseveld HW, Stouthamer AH, Boogerd FC (1995) Genetic and phenetic analyses of *Bradyrhizobium* strains nodulating peanut (*Arachis hypogaea* L.) roots. Appl Environ Microbiol 61: 1599-1609.

Willems A, Collins MD(1993) Phylogenetic analysis of rhizobia and agrobacteria based on 16S rRNA gene sequences. Int J Syst Bacteriol 43:305-313.

Wong FYK, Stackebrandt E, Ladha JK, Fleischman DE, Date RA, Fuerst JA (1994) Phylogenetic analysis of *Bradyrhizobium japonicum* and photosynthetic stem-nodulating bacteria from *Aeschynomene* species grown in separate geographical regions. Appl Environ Microbiol 60:940-946.

Yanagi M, Yamasato K(1993) Phylogenetic analysis of the family *Rhizobiaceae* and related bacteria by sequencing of 16S rRNA gene using PCR and DNA sequencer. FEMS Microbiol Lett 107:115-120.

Young JPW, Downer HL, Eardly BD (1991) Phylogeny of the phototrophic rhizobium strain BTAi1 by polymerase chain reaction-based sequencing of a 16S rRNA gene segment. J Bacteriol 173:2271-2277.

Comparison of Biolog and Phospholipid Fatty Acid Patterns to Detect Changes in Microbial Community

Ansa Palojärvi[1], S. Sharma[2], A. Rangger[2], M. von Lützow[1] and H. Insam[2]

[1] GSF - National Research Center for Environment and Health, Institute of Soil Ecology, D-85758 Neuherberg, Germany
[2] University of Innsbruck, Institute of Microbiology, A-6020 Innsbruck, Austria

Abstract. To determine phenotypic composition of microbial communities in soil, community level physiological profiles by Biolog® GN microtiter plates, and phospholipid fatty acid (PLFA) analysis were applied. An incubation experiment was carried out to compare the methods for soils of different origin (Denmark, Germany, Italy) with and without maize straw application. The straw was placed either on the soil surface (mulching) or incorporated into the soil. Destructive samplings were carried out 0, 2, 4, 16, and 52 weeks after the application. Profiles of utilization capacity of carbon substrates and PLFAs were statistically analyzed by two multivariate methods: principal component analysis (PCA) and discriminant analysis (DA). The Biolog and PLFA patterns often showed the same trend but PLFA achieved more clear groupings. Selective growth of bacteria adapted to the conditions in Biolog microtiter wells might explain some differences between the methods. The basic recommendations to apply PCA (e.g. sufficient observations to variables ratio) were not always fulfilled. Discriminant analysis was applicable for data sets since grouping was known *a priori*, and led to better separation than PCA.

Keywords. Microbial community, Biolog, phospholipid fatty acids, multivariate statistics, agricultural soil, maize straw, straw placement

1. Introduction

Recently, several methods have been suggested for soil microbial community analysis. Since less than 10 % of all microbes are culturable (Torsvik *et al.* 1994), conventional microbiological methods for description of communities must fail. DNA hybridization (Griffiths *et al.* 1996) and denaturing gradient gel electrophoresis (DGGE; Muyzer *et al.* 1993) are among the most promising approaches to describe genotypic structure of microbial communities. To determine phenotypic composition of the communities, substrate utilization profiles detected by Biolog microtiter plates, and fatty acid methyl ester (FAME) patterns have frequently been used lately.

Garland and Mills (1991) introduced use of community-level carbon source utilization patterns for comparison of microbial communities from different habitats. Similarly, Winding (1994) and Zak *et al.* (1994) analyzed Biolog GN substrate utilization profiles from samples taken from different environments. The Biolog approach gives information about substrate utilization and functional diversity (Zak *et al.* 1994) of soil bacteria. Insam *et al.* (1996) detected changes in functional abilities of microorganisms during composting of manure. Knight *et al.* (1997) found effects of both pH and heavy metal stress on the metabolic potential of soil microbes. The results from Garland (1996b) suggested that the Biolog approach is effective for detecting plant development differences in rhizosphere communities, and changes in response to plant developmental state.

The advantage of the Biolog approach is its easy applicability, making the method feasible for use in large-scale field studies. However, there are several aspects and drawbacks that must be taken into account. Standardization of inoculum density is important, since inoculum cell density and the rate of color development are strongly correlated (Garland and Mills 1991; Winding 1994). Results from Garland (1996a) suggest that single-plate readings can be used to classify samples, but only if potential differences in average well color development (AWCD) are accounted for in the data analysis. Additionally, the method reflects the metabolic capabilities of only a subset of the whole community since the method is selective for organisms actively metabolizing under the given conditions on the microtiter plates. These conditions usually deviate largely from those in the environment (Bossio and Scow 1995).

The methods to detect fatty acid methyl esters (FAME) are based on total extraction of lipids from a sample without separation of e.g. culturable and non-culturable bacteria. Results showing differences in whole community phospholipid fatty acid (PLFA) patterns due to e.g. agricultural management (Zelles *et al.* 1992), laboratory handling (Petersen and Klug 1994) and heavy metal treatments (Frostegård *et al.* 1993; 1996) have been published. Some authors (e.g. Haack *et al.* 1994; Cavigelli *et al.* 1995) have used the Microbial Identification Systems (MIDI or MIS) to detect whole cell fatty acid patterns. These methods were developed for taxonomical separation of bacterial isolates grown on standardized media. In these methods all lipids are included. The advantage to analyze the phospholipids (PLs) only is that they are located in membranes of all living cells, but not in storage lipids. Different subsets of a microbial community contain different phospholipid fatty acids in their membranes, or at least different compositions of them. Additionally, PLs have a relatively short turnover time (review by Tunlid and White 1992).

The microbial community characterization by Biolog and PLFA methods is based on specific profiles of either utilization capacity of carbon substrates or PLFAs. To differentiate soil microbial communities by carbon source metabolism or PLFA patterns, several multivariate statistical methods have been used. The analysis method should be chosen according to the specific question in mind. The most commonly used approach is the principal component analysis (PCA). PCA includes as much variation of original variables as possible to the new few principal components, which are independent from each other. Discriminant analysis (DA) is a method to analyze differences between known groups. The analysis searches for those variables that separate the groups at their best. The stepwise selection procedure adds variables one by one to the analysis until no further increase in the separation power is achievable (Backhaus *et al.* 1990).

So far, comparative studies on different methods describing microbial communities have been lacking. As suggested by Garland (1996b), comparative studies between C source profiling and other community level methods such as phospholipid fatty acids should be carried out to evaluate the utility of Biolog approach.

The aim of this study was to compare Biolog and PLFA methods to detect differences in microbial community structure (1) in soils of different origin and (2) after different straw applications (mulching *vs.* incorporation). Two different statistical approaches, (a) principal component analysis and (b) discriminant analysis, were applied.

2. Materials and Methods

2.1 Laboratory incubation

Table 1 represents the main soil chemical and physical parameters of the three soils used for the experiment. The soils were originating from Denmark (DK), Germany (DE) and Italy (IT), sampled near the cities of Roskilde, Munich and Naples, respectively. The management system of the DK site was a low input system growing barley in 1994. The DE site was a part of the experimental farm of the FAM network (Forschungsverbund Agrarökosysteme München). In 1994 the field had been cropped with potatoes. The IT site was a previous vineyard and had been fallow for 20 years.

Table 1. Main soil chemical and physical parameters of the soils from Denmark (DK), Germany (DE), and Italy (IT)

Site	pH	Texture [%]			Bulk density [g cm^{-3}]	C_{org} [%]	N_t [%]
		Sand	Silt	Clay			
DK	6.6	47	43	10	1.4	1.52	0.150
DE	6.4	17	62	22	1.5	1.26	0.145
IT	6.3	37	39	24	1.5	1.24	0.126

Soil samples were taken in September 1994 from the topsoil. The soils were air dried and sieved (2 mm). The soil columns used for the incubation were 7 cm high with 10 cm diameter. Each column was filled with 700 g rewetted soil (594 g dw, 584 g dw, and 586 g dw for Danish, German, and Italian soils, respectively). The maize straw used (leaves and stems) was passed through a 2 cm sieve. The initial total C and N % values of the straw were 45.3 and 0.5, respectively, giving a C/N ratio of 90.5. One third of the columns remained untreated without straw application (control), and the following two treatments were conducted: (1) maize litter placed on the soil surface (mulch treatment) and (2) maize incorporated into the soil. Straw addition corresponded to 10 t ha^{-1} (7.8 g straw 78 cm^{-2}). The incubation was carried out at 14°C and 40 - 60 % of the maximal water holding capacity (mWHC). Humidity was controlled by weighing twice a week. For the mulch treatment an upper layer sample (straw layer together with 0.5 cm soil) and a lower layer sample (0.5-7cm) were analyzed separately. Samples were taken before, and 2, 4, 16, and 52 weeks after the straw application, each time in triplicate.

2.2 Community level physiological profile by Biolog®

Bacterial suspensions were obtained by mixing 1 part of soil with 9 parts of NaCl (0.85%) (w/w) and shaken for 1 h at 250 rpm. To remove larger particles, suspensions were allowed to settle for 1 h. The supernatant was used for further dilution. Samples were diluted as far as necessary to achieve negligible background color and to obtain an approximate cell density (acridine orange direct count) of $3x10^5$ ml^{-1}. Each well of Biolog GN plates was inoculated with 125 µl of this suspension. Three replicate plates for each sample were incubated at 14°C. Absorbance was read at 592 nm with a microtiter plate reader (SLT SPECTRA; Grödig, Austria). First absorbance readings were made after significant color development was observed, this was usually after 96 h of incubation. The plates were examined twice a day. To avoid possible confounding effects of inoculum density, a standardization was made to select an appropriate time of reading for each plate. That point of time was selected when the average value of all wells in a plate was 1.1 ± 0.1 or when at least 10 wells had reached an absorbance value > 2.

Mulch litter (upper layer) and mulch soil (lower layer) were analyzed separately. Additionally, a mixture of mulch litter and mulch soil in representative proportions was examined to obtain the results for the mulch treatment as a whole. The data were transformed by calculating average well color development (AWCD) and then dividing the values for individual wells by AWCD of the plate (Garland and Mills 1991), modified by Insam et al. (1996). The transformed data were used for statistical analysis.

2.3 Phospholipid fatty acids (PLFA)

Phospholipid fatty acid patterns were examined according to Frostegård et al. (1993), with slight modifications. All chemicals used were of analytical grade. Glassware was washed carefully and ignited for 4 h at 400°C before use to avoid lipid contamination. Lipid extraction and fractionation was carried out from 2 g (wet weight; ww) of soil or 0.5 g (ww) of straw. Lipids were extracted from the samples using a one-phase mixture of chloroform, methanol and citrate buffer (0.15 M, pH 4). The fraction was stored at -20°C until further fractionation. The lipids were separated into neutral, glyco- and phospholipids on columns containing silicic acid (Varian Bond Elut SILICA Si 500 mg) by eluting with chloroform, acetone and methanol, respectively. Methyl nonadeconoate was added to the phospholipid fractions as an internal standard. Following a mild, alkaline hydrolysis of phospholipids, the resulting fatty acid methyl esters, dissolved in isooctane, were separated and quantified by gas chromatography (Varian) equipped with a flame-ionization detector and a 50 m long non-polar phenylmethyl silicone capillary column (Hewlett-Packard). Helium was used as carrier gas. The injector and the detector had a temperature of 280°C and 300°C, respectively. The time-temperature program for the oven was as following: initial temperature 70°C for 2 min, increase 30°C min^{-1} until 160°C, increase 3°C min^{-1} until 280°C, final temperature 280°C for 15 min. Identification of the fatty acid methyl esters was done using mass spectrometry (Hewlett-Packard GC/MS system). All together, 36 PLFAs were identified. The individual PLFA values were transformed to mol%. Mulch litter and mulch soil were analyzed separately. The results for the whole mulch treatment were calculated from the separate values using the actual proportions of each sampling time.

2.4 Statistics

We used transformed (see earlier) data for PCA and discriminant analysis (DA). DA was carried out with forward stepwise selection. Predicted group membership more than 85% was considered successful. SPSS for Windows was used for statistical analysis.

3. Results

3.1 Origin of soil

When Biolog data were pooled for control soils from all sampling dates, PCA did not yield any grouping according to the origin of soils (Fig. 1 A). Variances of the Danish control samples were low, but the samples were not separated from the other control soils. Eight out of 95 substrates had high loading (> |0.8|) for PC1. From PLFA profiles, the PCA separated the Italian soil from other soils (Fig. 1 B) with seven out of 36 PLFAs with high loading for the first PC. It is possible that Biolog and PLFA profiles were not stable over incubation time. Therefore pooling of different sampling dates might not be appropriate for the characterization of soil origin. Additionally, only 40 - 50 % of all variation was explained by the first two principal components. Restricting analysis to the data of the two earliest sampling dates (0 and 2 weeks after the start of the incubation), the amount of variation explained by the PC1 and PC2 increased to >55 %, and a clear grouping according to the origin of soils was found (Figs. 2 A, B). Additionally, more substrates and PLFAs were specifically involved in the separation (29 and 11 variables, respectively, having high loadings by PC1).

When grouping of samples is known *a priori* (e.g. different countries), discriminant analysis can be carried out. Taken the same Biolog data set as for

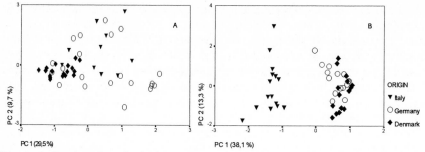

Fig. 1. Ordination plot of principal component analysis of Danish, German and Italian control soil samples according to (A) Biolog and (B) PLFA patterns. Pooled data over the whole incubation period (5 sampling dates, n=3).

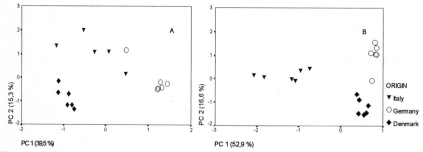

Fig. 2. Ordination plot of principal component analysis of Danish, German and Italian control soil samples according to (A) Biolog and (B) PLFA patterns. Pooled data from sampling dates 0 and 2 weeks after the straw application (n=3).

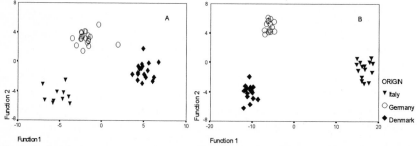

Fig. 3. Ordination plot of discriminant analysis of Danish, German and Italian control soil samples according to (A) Biolog and (B) PLFA patterns. Pooled data over the whole incubation period (5 sampling dates, n=3).

PCA in Fig. 1 A, the discriminant analysis was able to select substrates according to which the group separation for the data set over the whole incubation time was good (Fig. 3 A). From the PLFA data, DA resulted in a similar separation pattern

(Fig. 3 B). For both analyses and all countries the predicted group membership was 100%.

When comparing PCA and DA of control samples, the number of variables having high loading and of those included to discriminant functions was equal. However, only 2 similar substrates and 5 similar PLFAs were found to be related to the first PCA and DA axes. The particular substrates and PLFAs most closely related to PC1 and PC2, as well as to discriminant functions will be reported elsewhere.

3.2 Treatments

When pooling the whole data set from all treatments, countries and sampling times, PCA from Biolog patterns did not show any grouping, neither according to treatments (Fig. 4 A) nor to origin of soils. PLFA patterns, analyzed by PCA (Fig. 4 B), clearly separated the mulch litter samples from all other samples. In this analysis, the PLFA 18:2w6, specific to eucaryotic cells, was excluded. PLFA patterns of the control and the soil layer under the straw in mulch treatment were found to be identical. When PLFA data were labeled according to the origin of soil, the Italian samples were separated from the Danish and German samples in control, incorporated and mulch treatments (figure not shown; see Fig. 4 B: the sample group on the right lower corner). Only in mulch litter samples, the origin of the soil samples was insignificant. By Biolog data, only 2 substrates had specific and high loadings for PC1, whereas by PLFA analysis 11 phospholipids were involved.

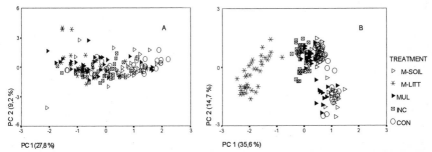

Fig. 4. Ordination plot of principal component analysis of control (CON), straw incorporated (INC) and straw mulched (MUL) treatments, and separately upper (litter) layer (M-LITT) and lower (soil) layer (M-SOIL) of mulch treatment according to (A) Biolog and (B) PLFA patterns. Pooled data from all countries over the whole incubation period (5 sampling dates, n=3).

Discriminant analysis did not show very clear groupings from Biolog data, when the whole data set was analyzed together and treatments were given as the basis for separation (Fig. 5 A). For all treatments the predicted group membership was < 80%. Control and mulch soils were similar to each other, and so were mulch treatment as a whole and the mulch litter. For PLFA patterns, the discriminant analysis was able to separate most of the treatments (Fig. 5 B). Litter layer from mulch treatment was clearly separated from the other treatments. For the whole mulch treatment, the predicted group membership of the samples was 94.5%. The predicted group membership of the samples where straw was incorporated was 88.9%. Control soils and soils under the mulch litter layer were not separated from each other. PLFA patterns in mulch treatment were very different from the patterns detected from mulch litter samples. DA included 9 and 8 variables from Biolog and PLFA data, respectively, to the discriminant functions. Only 1 and 2 of the same substrates and PLFAs, respectively, had high loadings by PC1.

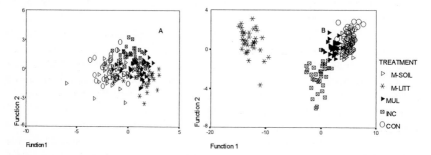

Fig. 5. Ordination plot of discriminant analysis of control (CON), straw incorporated (INC) and straw mulched (MUL) treatments, and separately upper (litter) layer (M-LITT) and lower (soil) layer (M-SOIL) of mulch treatment according to (A) Biolog and (B) PLFA patterns. Pooled data from all countries over the whole incubation period (5 sampling dates, n=3).

4. Discussion

4.1 Method comparison

Biolog microtiter plates have widely been used to detect the community level physiological profiles to describe the functional structure of microbial communities in different environments or after certain treatments (e.g. Garland and Mills 1991; Winding 1994; Zak *et al.* 1994; Insam *et al.* 1996; Knight *et al.* 1997). Haack *et al.* (1995) concluded from model microbial communities producing unique Biolog C utilization profiles that the method is capable to reflect real differences in the community. On the other hand, the metabolic capabilities of only a subset of the whole community can be reflected since the method is selective for bacteria actively metabolizing under the given conditions on the microtiter plates (Bossio and Scow 1995). The phospholipid fatty acid (PLFA) method is based on total extraction of PLFAs from the whole microbial community without separation of e.g. culturable and non-culturable organisms. The results from sediments (White *et al.* 1979), and soils with substrate additions

(Albers *et al.* 1994) indicate that rapid changes in microbial community structure can be detected by change in PLFA patterns.

In our study, substrate utilization patterns by Biolog and PLFA profiles showed often similar patterns, but in most cases Biolog did not give as clear groupings as PLFA. Both methods were able to differentiate the control soil samples according to the origin of soils. Biolog data showed no clear differences between the treatments when the whole data set was pooled. Very similar were the control soils and the soil layer in mulch treatment, as well as mulch treatment and the samples from upper straw layer of the mulch treatment. Confirming the Biolog results, control soils and lower soil layer in mulch treatment were identical according to PLFA profiles. However, all the other sample groups could also be separated by PLFA. Opposite to Biolog data, mulch treatment as a whole and the straw layer samples were very different. Garland and Mills (1991) and Winding (1994) noted that bacterial growth occurred in Biolog microplate wells. Both suggested that the Biolog assay reflects only that part of the community which is capable of active metabolism under the given conditions. In our experiment, the subset of microorganisms growing on the surface of maize straw might have been adapted to carbon-rich environment and so could rapidly grow in Biolog wells and outcompete the slower growing soil microorganisms. Another reason for deviating results could have been that PLFA analysis might be biased by some specific eukaryotic PLFAs from maize straw. That is why care was taken that the PLFAs possibly detectable from the straw itself (like 18:2w6; indicator for fungal and other eucaryotic cells) were excluded from the discriminant analysis. Also, the PLFAs found at concentrations close to the detection limit were not allowed to act as 'discriminating' PLFAs. The remaining PLFAs are not strictly confined to the bacterial members of the community (as the Biolog test is), but also fungal and actinomycete populations are measured. Important differences between the communities of unamended and straw-amended soils, and also between the methods, may actually be found especially with respect to these two groups.

4.2 Statistical analysis

The most often applied multivariate analysis for substrate utilization and PLFA patterns has been PCA (e.g. see Haack *et al.* 1994; Garland 1996a; Insam *et al.* 1996; Knight *et al.* 1997; for Biolog, and Frostegård *et al.* 1993; 1996; Petersen and Klug 1994; Haack *et al.* 1994; Cavigelli *et al.* 1995 for PLFA). The statistical method for Biolog and PLFA profile analysis should be chosen according to the specific question in mind. PCA includes as much variation of original variables as possible to the new few principal components, which are independent from each other. It is recommended that number of observations relative to variables should be 3:1 in PCA (e.g. Jackson 1993) or at least there should be more observations than variables. This is usually not the case especially in Biolog data, where there are 95 different substrates as variables. Discriminant analysis (DA) is a method to analyze differences between known groups. Since in our study the group membership was known *a priori*, DA was applicable. DA selects those variables that yield the best separation according to the given groups.

After PCA, scores for variable loadings can be evaluated to determine which variables provide the best separation power. In PLFA data, the number of variables with high loading (>|0.8|) was always higher than in Biolog data. Similarly to variables having high loadings in PCA, the variables selected by DA can be seen as the most important variables for the group separation. However, when comparing the variables having high loading in PCA with the ones selected by DA, the similarity was low. This may be partly due to the fact that the amount of variation explained by the first PCs was mostly less than 50% of the total variation.

The DA suggested that there are certain subsets of substrates and PLFAs specific to the origin of soil, partly overlain by changes during the 1-year incubation. Similarly, a subset of PLFAs reflected the different straw amendments. When comparing PCA from pooled control samples with the whole pooled data set, Biolog data had no and PLFA data had 4 identical variables with high loadings. The same comparison from DA shows that both Biolog and PLFA data had one identical variable with high loading.

In Biolog, 95 substrates might be too many for proper statistical analysis. In addition to the statistical problems resulting in often unfavourable observations to variables ratio, it is reasonable to assume that not all members of the microbial community react to the treatment under consideration and, moreover, not all substrates are relevant for the grouping. In rather heterogenous data, or in data sets containing many variables not influenced by the effect studied (e.g. changes in utilization patterns of only few C sources) the first 1-3 PCs in PCA might explain less than 50 % of the total variance of the data set. This makes it difficult to interpret the results. Compared to the 95 substrates of Biolog plates, in PLFA method we had only 36 variables, which made PCA analysis more reliable.

In this study, a data set from a multidisciplinary project was used to compare two different methods of microbial community characterization. Additionally, two statistical multivariate analyses were carried out. The data set will further be analyzed, and the results will be discussed in a broader context together with other parameters related to the decomposition of maize straw in soils of different climatic origin. The results presented in this paper suggest that the Biolog and PLFA patterns have often the same trend, but at least in pooled data sets PLFA achieved more clear groupings. Growth of only a subset of bacteria better adapted to the conditions in Biolog microtiter wells might explain some differences between the methods.

For future studies it is recommended to largely increase the observations-to-variables ratio, especially for the Biolog assay. For this purpose Biolog EcoPlates are now available, containing only 31 substrates in 3 replications, which were selected based on work by Campbell *et al.* (1997) and Hitzl *et al.* (1997).

Acknowledgements. We thank Juliane Filser and Peter Dörsch for helpful criticism, and Gudrun Hufnagel and Elke Bartusel for taking care of the incubation experiment. This study was financed by the Environment Program of the European Union (research project MICS; EV5V-CT94-0434) and by the research network "Forschungsverbund Agrarökosysteme München" (FAM). The scientific activities of the FAM are financially supported by the Federal Ministry of Research and Technology (BMBF 0339370). Rent and operating expenses are paid by the Bavarian State Ministry for Education and Culture, Science and Art. Shobha Sharma was supported by the Austrian Academic Exchange Service (ÖAD).

5. References

Albers B, Zelles L, Bai Q, Lörinci G, Hartman A, Beese F (1994) Fettsäuremuster von Phospholipiden und Lipopolysacchariden als Indikatoren für die Struktur von Mikroorganismengesellschaften in Böden. In: Proceedings: Eco-Informa -94, 5.-9.9.1994, Band 5, Umweltmonitoring und Bioindikation

Backhaus K, Erichson B, Plinke W, Weiber R (1990) Multivariate Analysemethoden. Springer-Verlag Berlin

Bossio DA, Scow KM (1995) Impact of carbon and flooding on the metabolic diversity of microbial communities in soils. Appl Environ Microbiol 61: 4043-4050

Campell C, Grayston SJ, Hirst D (1997) Use of rhizosphere C sources in sole C source tests to discriminate soil microbial communities. J Microbial Methods (in press)

Cavigelli MA, Robertson GP, Klug MJ (1995) Fatty acid methyl ester (FAME) profiles as measures of soil microbial community structure. In: Collins P, Robertson GP, Klug MJ (eds) The significance and regulation of soil biodiversity, Kluwer Academic Publisher, the Netherlands, pp. 99-113

Frostegård Å, Tunlid A, Bååth E (1993) Phospholipid fatty acid composition, biomass, and activity of microbial communities from two soil types experimentally exposed to different heavy metals. Appl Environ Microbiol 59:3605-3617

Frostegård Å, Tunlid A, Bååth, E (1996) Changes in microbial community structure during long-term incubation in two soils experimentally contaminated with metals. Soil Biol Biochem 28: 55-63

Garland JL (1996a) Analytical approaches to the characterization of samples of microbial communities using patterns of potential C source utilization. Soil Biol Biochem 28: 213-221

Garland JL (1996b) Patterns of potential C source utilization by rhizosphere communities. Soil Biol Biochem 28: 223-230

Garland JL, Mills AL (1991) Classification and characterization of heterotrophic microbial communities on the basis of patterns of community-level sole-carbon-source utilization. Appl Environ Microbiol 57: 2351-2359

Griffiths BS, Ritz K, Glover LA (1996) Broad-scale approaches to the determination of soil microbial community structure: application of the community DNA hybridization technique. Microbial Ecology 31: 269-280

Haack SK, Garchow H, Odelson DA, Forney LJ, Klug MJ (1994) Accuracy, reproducibility, and interpretation of fatty acid methyl ester profiles of model bacterial communities. Appl Environ Microbiol 60: 2483-2493

Haack SK, Garchow H, Klug MJ, Forney LJ (1995) Analysis of factors affecting the accuracy, reproducibility, and interpretation of microbial community carbon source utilization patterns. Appl Environ Microbiol 61: 1458-1468

Hitzl W, Rangger A, Sharma S, Insam H (1997) Separation power of the 95 substrates of the Biolog system determined in various soils. FEMS Microbiol Ecol (in press)

Insam H, Amor K, Renner M, Crepaz C (1996) Changes in functional abilities of the microbial community during composting of manure. Microbial Ecology 31:77-87

Jackson DA (1993) Stopping rules in principal components analysis: a comparison of heuristical and statistical approaches. Ecology 74: 2204-2214

Knight BP, McGrath SP, Chaudri AM (1997) Biomass carbon measurements and substrate utilization patterns of microbial populations from soils amended with cadmium, copper, or zinc. Appl Environ Microbiol 63: 39-43

Muyzer G, De Waal EC, Uitterlinden AG (1993) Profiling of complex microbial populations by denaturing gradient gel electrophoresis analysis of polymerase chain reaction-amplified genes coding for 16S rRNA. Appl Environ Microbiol 59: 695-700

Petersen SO, Klug MJ (1994) Effects of sieving, storage, and incubation temperature on the phospholipid fatty acid profile of a soil microbial community. Appl Environ Microbiol. 60:2421-2430

Torsvik V, Goksoyr J, Daae FL, Sorheim R, Michalsen J, Salte K (1994) Use of DNA analysis to determine the diversity of microbial communities. In: Ritz K, Dighton J, Giller KE (eds) Beyond the Biomass: compositional and functional analysis of soil microbial communities, John Wiley & Sons Ltd., Chichester, United Kingdom, pp. 39-48

Tunlid A, White DC (1992) Biochemical analysis of biomass, community structure, nutritional status, and metabolic activity of microbial communities in soil. In: Stotzky G, Bollag J-M (eds) Soil Biochemistry , Vol.7, Marcel Dekker, New York, pp. 229-262

White DC, Davies WM, Nickels JS, King JD, Bobbie RJ (1979) Determination of the sedimentary microbial biomass extractable lipid phosphate. Oecologia 40:51-62

Winding AK (1994) Fingerprinting bacterial soil communities using Biolog microtiter plates. In: Ritz K, Dighton J, Giller KE (eds) Beyond the Biomass: compositional and functional analysis of soil microbial communities, John Wiley & Sons Ltd., Chichester, United Kingdom, pp. 85-94

Zak JC, Willig MR, Moorehead DL, Wildman HG (1994) Functional diversity of microbial communities: a quantitative approach. Soil Biol Biochem 26: 1101-1108

Zelles L, Bai Q, Beese F (1992) Signature fatty acids in phospholipids and lipopolysaccharides as indicators of microbial biomass and community structure in agricultural soils. Soil Biol Biochem 24: 317-323

Combined Application of Biolog and MIS/SHERLOCK for Identifying Bacterial Isolates from Hydrocarbon-polluted Argentinian Soils

Lothar Wünsche[1], C. Härtig[1], H. O. Pucci[2] and W. Babel[1]

[1] UFZ-Umweltforschungszentrum Leipzig-Halle GmbH, Sektion Umweltmikrobiologie, Postfach 2, 04301 Leipzig, Germany; [2] C.E.I.M.A. Universidad Nacional de la Patagonia SJB, 9000 Comodoro Rivadavia, Argentina

Abstract. Taxonomic diversity and the hydrocarbon-degrading potential of heterotrophic, aerobic biocenoses of three different ecosystems (mineral oil-contaminated, unpolluted and bioremediated soils) were assessed by characterizing and identifying their main bacterial constituents. All colonies grown on universal (R2A agar) and selective (mineral agar with different mineral oils as the sole C-source) media on plates with suitable colony densities were isolated. The automated identification systems BIOLOG and MIS/SHERLOCK were used for the rapid identification of these isolates. In each case, the assignment to a taxon was verified by determining further diagnostically relevant parameters (cell and colony morphology, selected biochemical reactions, chemotaxonomical criteria). Corresponding to earlier results, the BIOLOG system was found to provide satisfactory results in the identification of Gram-bacteria whereas within the group of Gram+ isolates only a few taxa were identified reliably. MIS/SHERLOCK provided sufficient results for a wide range both of Gram- and Gram+ isolates. The occurrence of characteristic fatty acids even allowed in some cases reliable taxonomical assignment to a genus despite relatively low similarity indices. However, some apparently reliable identification results obtained with both identification systems proved to be incorrect according to additional diagnostic methods applied. The combined use of BIOLOG and MIS/SHERLOCK in connection with some easily accessible and routinely determined diagnostically relevant properties resulted in rates of correct identification of approximately 75% at genus level and and about 30 to 60% at species level (compared to the total number of isolates). Because of the very different identification results within certain taxonomical and physiological groups, the percentage of identifiable bacterial components of terrestrial biocenoses varied with their taxonomic composition.

Keywords. Identification of soil bacteria, BIOLOG system, MIS/SHERLOCK system, substrate utilization pattern, fatty acid profile

1. Introduction

Knowledge of the taxonomic and physiological diversity of the autochthonous biocenoses of natural ecosystems can provide insights into the ecological function of these communities (Atlas, 1984) and must be regarded as an essential presupposition for decisions concerning important problems of applied environmental microbiology (Table 1). A very wide range of methods is available for the identification of microorganisms. Due to the specific pros and cons of each method connected with the time, material and equipment required, selecting the identification method to be used depends on the input necessary and the quality of results desired.

Table 1. Importance of analyzing the taxonomic and physiological diversity of autochthonous biocenoses of natural ecosystems for applied environmental microbiology

Microbial-ecological investigation	Importance to applied environmental microbiology/bioremediation
Distribution of microbial taxa and their metabolic potential over natural ecosystems	Assessment of the applicability of bioremediation of a polluted ecosystem
	Choice of the method of bioremediation, e.g. in-situ- bioremediation using the (selectively supported) autochthonous biocenoses, application of starter cultures
	Isolation of special strains with high degrading potential adapted to the ecological situation in the relevant ecosystems
	Occurrence of taxa of risk groups ≥ 2
Variability of microbial autochthonous biocenoses as function of changing biotic and abiotic environmental factors	Estimation of the influence of allochthonous wild strains/GEM's (released unintentionally or with an aim) on structure and function of autochthonous biocenoses
	Evaluation of the ecological situation of an ecosystem (bioindication)
	Biomonitoring of pollution and bioremediation
Fate of allochthonous microorganisms in natural ecosystems	Survival/propagation of wild strains or GEM's (released unintentionally or with an aim) in natural ecosystems

In principle, commercially available automated identification systems offer favourable possibilities for identifying isolates from environmental samples. These systems were originally developed for the identification of clinically important microorganisms with well-defined properties. Only a few reports, however, exist about the application of automated systems to identify environmental isolates with mostly unknown properties. Klingler et al. (1992) identified by using the BIOLOG system 93% of 45 unknown Gram- isolates from the prototype of a water recycling system proposed for use on the U.S. space station Freedom. With the same method, Wünsche and Babel (1996) were able to identify 70% of the Gram- but only 35% of the Gram+ isolates from terrestrial habitats. Amy et al. (1992) applied the systems API-NFT (bioMerieux, France), BIOLOG (Biolog Inc., U.S.A.) and MIS (Microbial ID Inc., U.S.A.) to the identification of bacterial environmental isolates and found that each system identified only a small percentage of the total isolates. Identical results found by more than one system were the exception rather than the rule.

Commercially available automated identification systems include different spectra of microorganisms. They are based on the measurement of different phenotypic properties, e.g. the utilization of diagnostically relevant substrates (BIOLOG system) or the fatty acid composition of the cell matter (MIS/SHERLOCK system). We used both systems to analyze the diversity of autochthonous bacterial biocenoses of Argentinian soils with three degrees of hydrocarbon pollution: contaminated by crude petroleum after the rupture of a pipeline, redeveloped by in-situ bioremediation, and unpolluted (reference). The degradation of hydrocarbon pollutions is mainly based on the metabolic activities of heterotrophic, aerobic or facultatively anaerobic bacteria. Therefore, the relevant part of the biocenoses involved ought in principle to be identified by both identification systems, which chiefly comprise taxa of these ecotypes.

The aim of the present study was to answer three questions:
- Does the simultaneous application of BIOLOG and MIS/SHERLOCK significantly extend the range of identifiable environmental isolates?
- Can the higher reliability of the identification results be expected?
- Are the identification results obtained plausible, i.e. do they agree with fundamental morphological and physiological- biochemical criteria of bacterial diagnostics?

2. Materials and Methods

2.1 Soil samples

The origin, treatment and some relevant properties of the soils investigated are listed in Table 2. The chemical-analytical data were kindly provided by Dr. V. Riis (Centre for Environmental Research Ltd., Leipzig-Halle).

2.2 Isolation of aerobic/facultatively anaerobic heterotrophic bacteria from the soil samples

The bacteria were separated from the soil matrix by gentle shaking of 10 g soil (with known content of dry matter) with 100 ml 0.2% solution of $Na_4P_2O_7$ for 30 minutes and following decantation of the supernatant suspension (DECHEMA guidelines, 1992). The suspension was diluted in suitable steps with saline and spread onto the surface of nutrient plates with R2A agar after Reasoner and Geldreich (1985) as well as mineral salt agar with hydrocarbons (50% crude petroleum and 50% gas oil) as the sole carbon source. After determining the number of culturable bacteria on both substrate types, all the colonies grown on plates with suitable colony density were isolated and identified.

Table 2. Selected characteristics of three Argentinian soils (oil field Cañadon Seco, Patagonia)

Characteristics	Sample		
	Contaminated by mineral oil	Unpolluted	In-situ-bioremediated
Dry matter (%)	75.8	85.0	69.1
Total hydrocarbon content (g kg^{-1} soil dry matter)	92.01	0.06	11.65
Content of inorganic nutrients (mg kg^{-1} soil dry matter)			
Ammonium-N	n.d.	< 1	4.9
Nitrate-N	n.d.	7.0	5.7
P	n.d.	118.0	248.3
Content of culturable bacteria[1]	2.1×10^7	6.3×10^6	1.1×10^8
Content of hydrocarbon utilizing bacteria[2]	1.1×10^7	2.3×10^6	3.0×10^6
% of the culturable bacteria, altogether	52.4	36.5	2.7

[1] colony forming units per g soil dry matter after cultivation on universal (R2A) agar
[2] colony forming units per g soil dry matter after cultivation on agar with mineral oil as sole C source
n.d. = not (exactly) determinable (due to the high content of petroleum)

2.3 Identification of the soil isolates

For the automated identification of the soil isolates the BIOLOG Automated Identification System, Microstation[TM], GN and GP data bases release 3.50 (Biolog, Inc., Hayward, Calif., U.S.A.) and the MIDI Microbial Identification System MIS/SHERLOCK, data bases for aerobic bacteria (TSBA standard library, version 3.9, 1995) and Actinomycetes (ACTIN1 standard library, version 3.8, 1993), Microbial ID Inc., Newark, Del., U.S.A., were used. The cultivation of the test strains, the preparation of the bacterial suspensions, the measurement of the properties diagnostically used and the evaluation of the results were carried out according to the instructions for the users of these systems.

The plausibility of the results obtained was tested by determining 'classical' diagnostically important morphological and physiological-biochemical characteristics as Gram behaviour, cell and colony morphology, motility, and, by using the API system (bioMerieux), nitrate reduction and the occurrence of some diagnostically relevant enzymes (oxidase, catalase, gelatinase, urease). In Gram+ isolates the occurrence of diaminopimelic acids according to Hasegawa et al. (1983) in combination with typical sugars (according to Becker et al., 1965) in the total cell hydrolysates was measured in order to separate and distinguish between the groups of *Actinomycetales*.

3. Results and Discussion

In order to assess the possibilities and limits of identifying environmental isolates merely by their substrate utilization patterns, the BIOLOG system was applied to the determination of the species affiliation of the morphologically distinguishable colonies isolated from three Argentinian soils with different degrees of mineral oil pollution (Table 3). These isolates represent the ecologically relevant main taxa (abundance $\geq 10^4$-10^5 cfu g^{-1} soil dry matter) of the bacterial biocenoses of these ecosystems.

In agreement with earlier results (Wünsche and Babel, 1996) it was found that a relatively high percentage of the Gram- soil isolates could be identified at species level (maximum 79%) while the identification in the group of Gram+ bacteria was inadequate (identification rate 0 to 15%). This contrast may be explained by the fact that important genera of soil bacteria are either not included in the data base of the Gram+ bacteria of the BIOLOG system (e.g. *Streptomyces* and other *Actinomycetales*) or only very sparsely (e.g. *Arthrobacter* with only one species). Consequently, the percentage of identifiable isolates varied with taxonomic composition, especially with the proportion of Gram+ taxa of the biocenoses investigated. If only definite identification results (similarity indices > 0.750 indicating excellent identification according to the BIOLOG manual) which have been additionally confirmed by some fundamental and routinely determined diagnostically relevant morphological and physiological-biochemical characteristics are taken into consideration, the percentage of reliably identifiable soil isolates decreased significantly.

The combined application of the commercially available identification systems BIOLOG and MIS/SHERLOCK, which are based on the determination of very different phenotypic characteristics (substrate utilization pattern and fatty acid profiles, respectively), extended the range of the reliably identifiable taxa. The additional measurement of some morphological, biochemical and chemotaxonomic parameters was found to be necessary to confirm the identification results obtained. Four examples illustrating the advantages of this approach are shown in Table 4. Compared with the BIOLOG system, the MIS/SHERLOCK system comprises a greater number of Gram+ taxa. Furthermore, the occurrence of distinct fatty acids (or combinations thereof), characteristic for greater groups or higher taxonomic categories, even allowed in some cases the assignment of an isolate to a genus despite relatively low similarity indices which made identification at species level doubtful.

Analyzing the taxonomic diversity of three Argentinian soils with different hydrocarbon contents by the described combination of methods (Table 5),

Table 3. Identification of soil isolates according to their substrate utilization patterns by using the BIOLOG system

	Sample		
	Contaminated by mineral oil	Unpolluted	In-situ-bioremediated
Number of morphologically distinguishable colonies	46	16	38
of them, Gram -	19	4	19
of them, Gram +	27	12	19
Number of isolates identifiable at species level (S $\geq$ 0.500)			
total	19	2	11
(% of the total colony number)	(41)	(13)	(29)
Gram - isolates	15	1	11
(% of the Gram - isolates)	(79)	(25)	(58)
Gram + isolates			
(% of the Gram + isolates)	4	1	0
	(15)	(8)	(0)

Table 5. Quantitatively predominating taxa within heterotrophic aerobic bacterial biocenoses of Argentinian soils (Patagonia, Cañadon Seco)

	Soil sample	
Highly contaminated by mineral oil	Unpolluted	In-situ-bioremediated
Acinetobacetr spec. *	*Arthrobacter spec.* *	*Alcaligenes xylosoxidans*
Arthrobacter spec.	*A. ramosus*	*Nocardia spec.* *
A. globiformis	*Cellulomonas turbata*	*N. restricta* *
A. oxidans	*Micrococcus luteus*	*Ochrobactrum anthropi* *
A. pascens	*Nocardia spec.* *	*Paenibacillus pabuli* *
Aureobacterium liquefaciens	*Pseudomonas spec.* *	*Pseudomonas fluorescens* *
Bacillus megaterium	*P. stutzeri* *	*P. putida* *
Micrococcus varians *	*Rhodococcus fascians*	*P. stutzeri* *
Nocardia spec.	*Streptomyces spec.* *	*P. vesicularis* *
N. asteroides		*Rhodococcus spec.* *
Pseudomonas spec. *		*Sphingobacterium spec.*
P. fluorescens *		*S. spiritivorum*
P. stutzeri *		*Streptomyces spec.* *
Rhizobium spec. *		*Xanthomonas spec.* *
Rhodococcus fascians *		
Streptomyces spec. *		
Not identifiable: 11 isolates = 24 %	Not identifiable: 4 isolates = 25 %	Not identifiable: 12 isolates = 32 %

* = Taxa with hydrocarbon utilizing isolates

Table 4. Examples of identifying bacterial soil isolates by the combined application of BIOLOG and MIS/SHERLOCK

Isolate	Identification by automated identification systems		Additional characteristics	Identification results
	BIOLOG	MIS/SHERLOCK		
AR 320	*Pseudomonas stutzeri* (similarity 0.779)	*Pseudomonas stutzeri* (similarity 0.446) *Pseudomonas syringae* (similarity 0.403)	Gram negative rods → section 4 Oxidase activity: Isolate AR 320: + *P. syringae*: − *P. stutzeri*: +	*Pseudomonas stutzeri*

AR 307	No identification	*Nocardia asteroides* (similarity 0.668)
		
		Actinomadura spec. (similarity 0.512)
			Gram positive, filaments fragmenting into shorter elements (rods)
			→ section 22
			DAP/diagnostic sugars:
			Isolate AR 307: m-DAP
			  arabinose, galactose
			Rhodococcus-Nocardia-group: m-DAP
			  arabinose, galactose
			Actinomadura: m-(or no)-DAP
			  madurose
				Nocardia asteroides

AE 157	No identification	*Streptomyces griseoflavus* (similarity 0.239)	Gram positive, aerial mycelium
			→ section 22
			DAP/diagnostic sugars:
			Isolate AE 157: L-DAP
			  galactose, glucose, ribose
			Streptomyces-group: L-DAP
			  not applicable
				Streptomyces spec.

AR 103	No identification	No identification	Gram negative according to Gram staining: Gram positive according to fatty acid profile short rods, no DAP	?

approximately 75% of the isolated soil bacteria could be assigned to a genus. Depending on the taxonomic composition of the analyzed biocenoses, especially on the proportion of Gram+ bacteria, about 30 to 60% of the isolates were identified at species level. Hydrocarbon-utilizing strains were found within the majority of the genera and species detected (these taxa were labelled in Table 5 with an asterisk). However, about 25 to 30% of the isolates from these different terrestrial environments could not be identified, i.e. the assignment to a genus (or a species) was too unreliable or the identification was not at all possible.

Four reasons responsible for the failure of the reliable identification of environmental isolates using the described methodological approach are proposed:

(1) The isolate belongs to a taxon which is not included in the data bases used.
(2) The isolate cannot grow (or only very slowly) under the standardized conditions (e.g. pH, temperature, medium composition, offered substrates) required by the identification systems.
(3) The isolate belongs to an included taxon but represents a strain with properties deviating from the information in the data bases (substrate utilization pattern or fatty acid profile under the given growth conditions).
(4) The isolate represents a new taxon not yet described which could be specific to the ecosystem being investigated.

4. References

Amy PS, Haldeman DL, Ringelberg D, Hall DH,Russell C (1992) Comparison of identification systems for classification of bacteria isolated from water and endolithic habitats within the deep subsurface. Appl Environ Microbiol 58: 3367-3373

Atlas R (1984) Diversity of microbial communities. Adv. Microb. Ecol. 7: 1-47

Becker B, Lechevalier MP, Lechevalier HA (1965) Chemical composition of cell wall preparations from strains of various form-genera of aerobic actinomycetes. Appl Microbiol 13: 236-243

DECHEMA guidelines (1992) Labormethoden zur Beurteilung der biologischen Bodensanierung. DECHEMA, Deutsche Gesellschaft für Chemisches Apparatewesen, Chemische Technik und Biotechnologie e.V., Frankfurt am Main

Hasegawa T, Takizawa M, Tanida S (1983) A rapid analysis for chemical grouping of aerobic actinomycetes. J Gen Appl Microbiol 29: 319-322

Klingler JM, Stowe RP, Obenhuber DC, Groves TO, Mishra SK, Pierson DL (1992) Evaluation of the BIOLOG Automated Microbial Identification System. Appl Environ Microbiol 58: 2089-2092

Reasoner DJ, Geldreich, EE (1985) A new medium for the enumeration and subculture of bacteria from potable water. Appl Environ Microbiol 49, 1-7

Wünsche L, Babel W (1996) The suitability of the BIOLOG Automated Microbial Identification System for assessing the taxonomical composition of terrestrial bacterial communities. Microbiol Res 151: 133-143

Diversity of Anthropogenically Influenced or Disturbed Soil Microbial Communities

Endre Laczko[1], A. Rudaz[2] and M. Aragno[3]

[1]Solvit, Langsägestrasse 15, CH-6010 Kriens; [2]IUL, Schwarzenburgstrasse 155, CH-3097 Bern; [3]Université Neuchâtel, rue Emile Argand 11, CH-2007 Neuchâtel

Abstract. The objectives of this study are to assess the influences of land use and of heavy metal loads on the soil microbial community. For this purpose we collected soils from 27 sites throughout Switzerland and analyzed saturated, mono- and polyunsaturated PLFAs (phospholipid fatty acids), ATP (total adenylates) and the total content of the heavy metal cadmium. We found a positive correlation between the sum of PLFA and ATP. Further, we found that land use and cadmium load are reflected by PLFA contents. Based on the organismic origin we formed guilds of individual PLFAs. These guilds relate to taxonomic units and trophic function at the same time. Therefore, it is possible to infer the structure (structural diversity) and trophic diversity (functional diversity) of the soil microbial communities. We found a negative correlation between functional diversity and cadmium load. Functional diversity differentiated also land use types. PLFA analysis may reveal structural and functional *in situ* properties of whole soil microbial communities. Substrate utilization tests, which are restricted to cultivable organisms, predominantly bacteria and fungi, may supplement PLFA analysis with data about specific activities. Finally, our results indicate that PLFA analysis can be used to estimate critical pollutant loads.

Keywords. Microbial community, structural diversity, functional diversity, soil, land use, heavy metals, cadmium, PLFA analysis, substrate utilisation

1. Introduction

The main goal of this study is to assess the influence of soil cadmium loads on the soil microbial community in a range of soils, which are representative for the intensively used Swiss midlands and for the extensively used remote areas.

A recent Swiss survey study revealed that topsoils often show cadmium, copper, lead and zinc loads which are above geogenic background contents (Vogel *et al.* 1989). These sites may be considered as polluted. The authors of the survey claim that this pollution has anthropogenic sources and affects the whole surface of Swiss soils and is still increasing. The results further indicate that Cadmium may serve as an indicator of direct and indirect anthropogenic impact. The survey showed that the heavy metal pollution has not yet reached levels known to reduce biomass or activity of the soil microbial community (e.g. Domsch 1985). The question may be raised, whether heavy metal loads above the geogenic background contents, but below the apparently toxic levels, already affect soil microbial communities. If heavy metal resistance is not the same among the species of a given community, then increasing heavy metal loads will exert a selective pressure and will alter community composition. If heavy metal sensitive microorganisms become eventually extinct, then this alteration may be accompanied by loss of metabolic capabilities. This loss must not immediately affect general activity or biomass of the community, because the loss might be compensated by unaffected species. A recent laboratory study of Burkhardt et al. (1993) on soil bacteria seems to support this view. The authors found that in bacterial communities affected by copper, nickel, lead or zinc, uncommon degradative capabilities are rarer than in unaffected ones. In a recent field study Pennanen et al. (1996) found changes in soil microbial community composition at soil copper contents that did not affect bacterial biomass, activity (respiration), and thymidine incorporation. Both studies deal with heavily polluted soils from the surroundings of metal mines or metal smelters. Similar investigations, which adress the functional or structural diversity of soil bacteria or microorganisms in soils, with moderate heavy metal load, do not exist, but would be highly desirable in establishing a sustainable soil conservation practice (Rudaz and Brüning 1990; Albers et al. 1994).

In our study the microbial communities are characterized in terms of biomass, diversity and trophic function. We infer all these parameters from the analysis of phospholipid fatty acids (PLFA), which are taxonomically relevant constituents of the cell membranes of all organisms. This method offers a quantitative approach towards assessment of the microbial community structure and minimizes methodological bias and artifacts (White 1995). PLFA analysis was already succesfully used by us to study the relation between soil microbial diversity and vitamin deficiency (Korner and Laczko 1992). Others used PLFA analysis for demonstrating the influence of land use practice (e.g. Zelles et al. 1995) or the effects of pollution (e.g. Baath 1992 et al.; Pennanen et al. 1996) on the soil microbial community.

With the present study it is shown how community structure and functional diversity may be inferred from the PLFA analysis and how functional diversity is related to structural diversity. Further the influence of land use and low-level cadmium pollution on the soil microbial community structure as well as on the trophic function within the community is shown. The second objective has general environmental relevance, because the addressed trophic functions are tightly connected with the cycling and mineralization of organic matter in soils (e.g. Hunt et al. 1987; Hassink et al. 1994).

2. Methods

For the purpose of the study we collected 27 soil samples in Switzerland. Four sampling sites are cultivated lands in a rural area of the Swiss midlands (area

1A). Three sites are cultivated lands in the vicinity of a steel factory and an urban area of the Swiss midlands (area 2A). Five sampling sites are cultivated lands in a rural area of the Swiss midlands, which were fertilized during several years with an industrial waste water sludge (area 3A). Seven sites are extensively used meadows in remote rural areas of the Swiss midlands, Jura mountains and prealpes (Areas 1M). Six sites are meadows in the vicinity of a steel factory and an urban area of the Swiss midlands (area 2M). Two sites are meadows in a rural area of the Swiss Jura mountains (area 3M). These last two sites are exclusively used for multidisciplinary research, conducted in the framework of the Priority Programme Environment, and are described in detail by Baur *et al.* (1996). According to Baur *et al.* (1996) these sites were intensively grazed by cows until 1993.

At each sampling site, 25 soil cylinders of not more than 20cm length were taken from the A horizon with a 30mm gauge auger and mixed to form one sample. The mixed sample was immediately sieved through a 6-mm mesh steel sieve. Visible plant roots were removed carefully. A homogeneous soil portion of about 12 to 15 g was immediately filled in a preweighed screw bottle containing the extraction solvents for PLFAs (see below). An equal portion for the determination of the moisture content of the freshly collected samples was filled in a preweighed gastight jar. The rest of the soil sample was stored at 4°C.

pH in water, the content of calcium carbonate, organic carbon and cadmium were analyzed according to the standard procedures of the Swiss Institute of Environmental Protection and Agriculture (FAC, 1989). Total adenylates were measured using the method of Maire (1982, 1984).

An extensive description of the PLFA analysis is given by Zelles *et al.* (1993). In short, we extracted the lipids from approximately 12 to 15 g of fresh soil (directly on the sampling site, see above) according to the procedure of Bligh and Dyer (1959). The phospholipids were isolated by silica gel chromatography and transesterified by a mild alkaline methanolysis. The methyl esters of the PLFA were then extracted and separated by two consecutive solid phase extractions into three fractions: saturated PLFA (SATFA), monounsaturated PLFA (MUFAS) and polyunsaturated PLFA (PUFAS). Each fraction was analyzed separately by GC-MS. The resulting chromatograms were evaluated by a software (Solvit E. Laczko, unpublished) able to identify and quantify 264 PLFAs on the basis of mass spectra, retention times and an internal standard.

We used the information about the organismic origin of the PLFA to form biological guilds of PLFA (in contrast to the chemical guilds like SATFA, MUFAS or PUFAS, or guilds defined by statistics like "PLFA showing covariation with heavy metal load"). The guilds we formed comprise PLFA from roots of higher plants, algae, protozoa, fungi, vesicular-arbuscular-mycorrhiza, cyanobacteria and bacteria (see appendix). Each of these guilds represents not only a taxonomic unit but also a distinct trophic function of the soil microbial community. Therefore, it is possible to infer the relative biomass of distinct taxonomic and functional units, and to characterize the microbial community structure, or to identify sensitive parts or functions within the community. Further, we applied an approach of Huston (1994) who suggested the use of a biologically meaningful diversity index and named it functional diversity. He defined functional diversity as the number of functional guilds multiplied by the mean number of units in those functional guilds. According to Huston (1994) a function may be a metabolic, an ecological or any other function which is related to a biotic structure. The units may be molecules, species or any other biotic structure. In our case we defined functional PLFA diversity as the number of PLFA guilds multiplied by the mean number of individual PLFAs in those guilds. To calculate functional diversity, the identified PLFAs of a sample were first classified into the mentioned guilds, then the number of PLFAs in each guild was determined (PLFA count). The relative biomass of a guild was estimated as the sum of the mass of the related PLFAs. PLFA diversity was calculated according to the proposition by Hill (1973). The diversity measure of Hill (1973) is used as a measure of the structural diversity of the soil microbial communities.

3. Results and Discussion

The sampling sites are briefly characterized in Table 1. All soil samples had a sandy loam texture (qualitative observation). The pH ranges from slightly acidic to neutral values. The organic carbon content is somewhat lower in cultivated lands and lies (with one exception) between 1.5% and 7% of the soil dry weight. All in all the analyzed soils may be considered similar in respect of texture, pH and organic carbon content.

Table 1. Characterization of the sampling areas

Use	Area (number of sites)	pH in water	organic C % of DW	Cd tot ng g^{-1} dry soil	Cd sol ng g^{-1} dry soil
Cultivated land	1A	6.4	1.6	255	1.5
	(4 sites)	(6.4-6.5)	(1.5-1.7)	(246-263)	(0.9-2.0)
	2A	5.6	1.1	204	4.8
	(3 sites)	(5.4-5.9)	(0.9-1.3)	(145-237)	(2.4-9.3)
	3A	6.4	1.6	416	14.8
	(5 sites)	(6.4-6.5)	(1.5-1.7)	(320-447)	(11.6-19.1)
Meadow	1M	6.3	6.6	240	<0.75
	(7 sites)	(4.1-7.6)	(1.1-23.5)	(72-481)	(below 0.75)
	2M	6.9	2.0	342	<0.75
	(6 sites)	(6.4-7.2)	(1.6-2.5)	(132-709)	(?-1.3)
	3M	7.3	2.5	1020	<0.75
	(2 sites)	(7.2-7.4)	(1.1-3.9)	(965-1074)	(below 0.75)

Values are means; values in brackets indicate the corresponding data range; tot = extracted with 1M nitric acid; sol = extracted with 0.1M aquous sodium nitrate solution

The selected sites are influenced by human activities in different ways. This is reflected by the soluble and total cadmium content shown in Table 1. In the rural and remote areas 1A and 1M a low anthropogenic impact may be suspected. The sole impacts are due to (similar) atmospheric depositions and land use practice. The pollution level may be considered low. In the areas 2A and 2M, in the vicinity of a steel factory, a point source of various pollutants, a gradient of low to high industrial impact may be suspected. The pollution level may be considered ranging from low to elevated. Although the total cadmium content is low in cultivated lands, the soluble cadmium content is relatively high. In the rural and remote areas 3A and 3M a very local and uniform agricultural impact may be suspected. The pollution level may be considered elevated or high. The somehow surprising cadmium pollution in the remote Jura mountains (area 3M) is not an isolated observation. Atteia *et al.* (1995) and the Office for Environmental Protection of the Canton Basel-Landschaft (Bono, 1993) found soil cadmium contents up to 1500 ng g^{-1} dry soil in the Jura mountains. The high soil cadmium contents were related to yet unknown pollution sources (Atteia *et al.* 1995).

The microbial characteristics of the analyzed soils are summarized in Table 2. The contents of all individual PLFAs were summed to obtain a measure of total PLFA. The obtained values for total PLFA correlate positively with total adenylates ($r = 0.812$, $n = 27$), indicating that we are analyzing PLFA from living biomass (Figure 1). Further, we observed a tendency towards low total PLFA contents (and low ATP contents) in cultivated soils and in soils with elevated cadmium loads (Figure 2). No relationship between ATP (biomass) and PLFA diversity or functional diversity was found (Figure 3). On the other hand we found negative correlations between the functional diversity and the total soil cadmium content of cultivated lands with cadmium loads above 200 ng g^{-1} dry soil ($r = -0.748$, $n = 12$) and of meadows with cadmium loads above 290 ng g^{-1} dry soil ($r = -0.904$, $n = 15$) (Figure 4). The mean PLFA mass of all PLFA guilds is lower in cultivated lands (areas 1A, 2A, 3A) and in soils with elevated cadmium loads (Table 2). The soil microbial communities of meadows seem to be more affected by the soil cadmium loads than the communities under cultivated lands. Different PLFA guilds are affected in different ways, indicating changes in the community structure (Table 2).

The presented results show that soil cadmium contents between 200 and 1000 ng g^{-1} dry soil have an influence on the soil microbial community. We found a weak negative correlation between soil cadmium load and microbial biomass. On the other hand we found no relationship between biomass and PLFA diversity or functional diversity. Thus the decrease of functional diversity at elevated soil cadmium contents reflects the impact of both land use and cadmium loads and is not an analytical artifact. We conclude that the microbial community is affected in meadows at cadmium contents above 300 ng g^{-1} dry soil and in cultivated lands above 200 ng g^{-1} dry soil. These values are 3 to 4 times lower than the pollution limits of the swiss legislation and 10 times lower than the levels which are reported to inhibit total soil respiration (e.g. Domsch 1985). Further, cultivated soils seem to have a lower critical load value of cadmium pollution than meadows. This means that any disturbance that reduces functional diversity permanently, as ploughing, results eventually in an enhanced susceptibility of the microbial community towards further disturbances. This is contrasted by the observation that the magnitude of biomass changes is bigger in meadows.

The impact of cadmium on the guilds is not significantly different. Nevertheless some tendencies were be observed. Regardless of the differences between cultivated lands and meadows, the biomass of bacteria and cyanobacteria covaries. The biomass of fungi and especially of protozoa seems to be more affected by cadmium loads than the biomass of bacteria or cyanobacteria. The opposite is true for the covarying biomass of plant roots and the associated VAM. Heavy metal sensitivity of fungi and heavy metal tolerance of VAM were also observed by Pennanen *et al.* (1996). Burkhardt *et al.* (1993) tested degradative capabilities of bacterial communities in the rhizosphere and found that heavy metal contamination reduced the ability of bacteria to degrade aromatic compounds.

The chemical analysis of structural cell components like PLFA delivers primarily structural information about the living biomass. Because we know the organismic origin of many PLFAs, we may relate structural information to taxonomic knowledge (e.g. Suzuki *et al.* 1993). The taxonomic relation finally

Table 2. Microbiological characteristics of the analyzed soils; the values are the means and the coefficients of variation (% of the means) for the sites in the areas characterized in Table 1

Area	ATP ng g^{-1}	PLFA nmol/g	FD	PLFA guilds (nmol/g) (values in brackets are counts of PLFAs in a guild)						
				R	A	P	F	VAM	CB	B
1A	972	36870	74	697	1391	1302	16960	2983	20909	33958
	12%	6%	3%	31%	28%	14%	13%	64%	4%	7%
				(1)	(3)	(3)	(14)	(3)	(19)	(31)
2A	1221	29290	70	526	956	531	10921	1753	14685	27155
	18%	13%	7%	44%	38%	3%	19%	128%	2%	16%
				(3)	(3)	(3)	(14)	(2)	(16)	(31)
3A	632	21950	60	295	636	589	10594	2112	12662	20816
	22%	25%	10%	71%	39%	65%	20%	60%	18%	24%
				(1)	(2)	(2)	(12)	(3)	(14)	(26)
1M	5524	126460	93	2530	6689	2345	64006	27075	55841	114702
	58%	40%	7%	69%	44%	32%	36%	70%	34%	40%
				(2)	(4)	(3)	(14)	(3)	(25)	(41)
2M	3562	71900	89	2151	3246	1250	30109	16735	28414	64659
	24%	27%	6%	41%	52%	26%	29%	46%	23%	28%
				(3)	(3)	(4)	(15)	(3)	(24)	(39)
3M	3273	35590	78	495	2834	698	17123	5228	17399	31982
	6%	0.4%	3%	121%	7%	3%	12%	28%	4%	2%
				(2)	(4)	(3)	(14)	(3)	(20)	(34)

FD = functional diversity; R = plant roots; A = Algea; P = Protozoa; F = Fungi; VAM = vesicular arbuscular Mykorrhiza; CB = Cyanobacteria; B = Bacteria

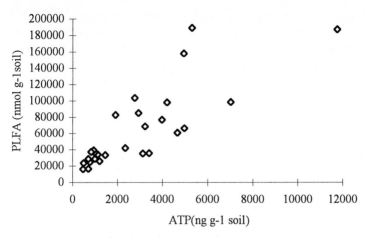

Figure 1. The relationship between total PLFA and ATP

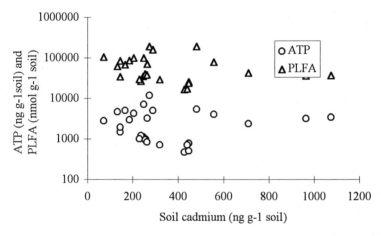

Figure 2. The relationship between PLFA, ATP and total soil cadmium content

allows to infer functional properties of the identified taxonomic units. In addition, the PLFA analysis provides structural information, which is nearly unbiased by methodological artifacts (White 1995). Thus PLFA analysis may reveal *in situ* structural and functional properties of the analysed community at the same time. However, most relationships between PLFA and taxonomic units, or functional properties, are not straight. Normally several interpretations of the PLFA data must be considered. This uncertainty was respected by the formation of the biological PLFA guilds. PLFAs with potentially multiple origin were related to all possible guilds. Substrate use analysis delivers primarily functional information about the living biomass, but structural information may be inferred, too (e.g. Dean-Ross 1989). In contrast to the PLFA analysis, substrate use must be tested always in an experimental approach, *ex situ*. The bias by methodological artifacts is strong and often not under control. Haack *et al.*

(1995) demonstrated for example the pitfalls and problems of the BIOLOG ® test. Moreover, such substrate use tests are, unlike PLFA analysis, restricted to bacteria. Nevertheless, substrate tests are the only way to confirm the presence of specific activities in microbial communities. In this way, substrate use tests may supplement structural analyses like the PLFA analysis and may reduce the uncertainty in the interpretation of structural data.

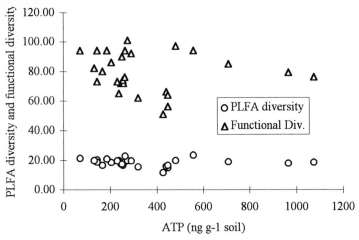

Figure 3. The relationship between PLFA diversity, functional diversity and ATP

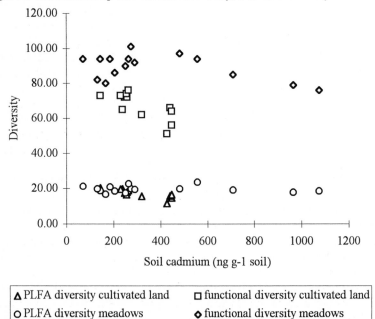

Figure 4. The relationship between PLFA (structural), or functional diversity and soil cadmium content

4. Conclusions

Biomass, community structure and functional diversity of soil microbial communities, as inferred from PLFA analysis, are interpretable parameters in low impact studies. Our study showed that moderate cadmium pollution reduces functional diversity of soil microbial communities. The relationship between functional diversity and pollutant load can be used to estimate critical pollutant loads. In the case of cadmium the critical loads, which affect the soil microbial community, are 290 ng g^{-1} dry soil in meadows and 200 ng g^{-1} dry soil in cultivated lands. This values are ten times lower than the values estimated on the basis of soil respiration tests. To clarify the impact of cadmium (or other pollutants) on specific guilds, or functions, of the soil microbial community, specific substrate use tests should be performed.

Acknowledgments. We thank Gertrud Fischer for measuring soil PLFAs, Astrid Schwarz for revising the presentation of the results and Karin Klapproth for revising the English. This study was supported by the Swiss National Science Foundation (grant 5001-34859) and Solvit.

5. References

Albers B, Zelles L, Bai Q, Lörinci G, Hartmann A, Beese F (1994) Fettsäuremuster von Phospholipiden und Lipopolysacchariden für die Struktur von Mikroorganismengesellschaften in Böden. In: Alef K, Fiedler H, Hutzinger O (eds) ECOINFORMA 5, Umweltbundesamt, Berlin, pp. 297-312

Atteia O, Thélin HR, Pfeifer, Dubois JP, Hunziker JC (1995) A search for the origin of cadmium in the soil of the Swiss Jura. Geoderma 68: 149-172

Baath E, Frostegard A, Fritze H (1992) Soil bacterial biomass, activity, phospholipid fatty acid pattern, and pH tolerance in an area polluted with alkaline dust deposition. Appl Environ Microbiol 58: 4026-4031

Baur B, Joshi J, Schmid B, Hänggi A, Borcard D, Stray I, Pedroli A, Thommen H, Luka H, Rusterholz H, Oggier P, Ledergerber S, Erhardt A (1996) Variation in species richness of plants and diverse groups of invertebrates in three calcareous grasslands of the Swiss Jura mountains. Revue suisse de Zoologie 103: 801-833

Bligh EG, Dyer WJ (1959) A rapid method of total lipid extraction and purification. Can J Biochem Physiol 37: 911-917

Bono R (1993) Schwermetalle in den Böden des Kantons Basel-Landschaft. Amt für Umweltschutz und Energie, Liestal.

Burkhardt C, Insam H, Hutchinson TC, Reber HH (1993) Impact of heavy metals on the degradative capabilities of soil bacterial communities. Biol Fertil Soils 16: 154-156

Dean-Ross D, Mills AL (1989) Bacterial community structure and function along a heavy metal gradient. Appl Environ Microbiol 55: 2002-2009

Domsch KH (1985) Funktionen und Belastbarkeit des Bodens aus der Sicht der Bodenmikrobiologie. Rat der Sachverständigen für Umweltfragen. Verlag W. Kohlhammer, Stuttgart und Mainz

FAC (1989) Methoden für Bodenuntersuchungen. Schriftenreihe der FAC Liebefeld Nr. 5, Liebefeld-Bern

Hassink J, Neutel AM, de Ruiter PC (1994) C and N mineralization in sandy and loamy grassland soils: the role of microbes and microfauna. Soil Biol Biochem 26: 1565-1571

Hill MO (1973) Diversity and evenness: a unifying notation and its consequences. Ecology 54: 427-432

Hunt HW, Coleman DC, Ingahm ER, Eliott ET, Moore IC, Rose SL, Reid CPP, Morley CR (1987) The detrital food web in a shortgrass prairie. Biol Fertil Soils 3: 57-68

Huston MA (1994) Biological Diversity. Cambridge University Press

Korner J, Laczko E (1992) A new method for assessing soil microorganisms diversity and evidence of vitamin deficiency in low diversity communities. Biol Fertil Soils 13: 58-60

Maire N (1982) Méthode de mesure de l'adénosine triphosphat (ATP) dans les sols. Bulletin de la Societé Suisse de Pédologie 6: 88-94

Maire N (1984) Extraction de l'adénosin triphosphat dans les sols: une nouvelle méthode de calcul des pertes en ATP. Soil Biol Biochem 16: 361-366

Olsson PA, Baath E, Jakobsen I, Söderström B (1995) The use of phospholipid and neutral fatty acids to estimate biomass of arbuscular mycorrhizal fungi in soil. Mycol Res 99: 623-629

Pennanen T, Frostegard A, Fitze H, Baath E (1996) Phospholipid fatty acid composition and heavy metal tolerance of soil microbial communities along two heavy metal-polluted gradients in coniferous forests. Appl Environ Microbiol 62: 420-428

Ratledge C, Wilkinson SG (1988) Microbial Lipids. Academic Press, London

Rudaz A, Brüning E (1990). Importance et contribution des microorgnismes pour l'élaboration de critères de fertilité des sols. Recherche agronomique en Suisse 29: 16-22

Suzuki K, Goodfellow M, O'Donnel AG (1993) Cell envelopes and classification. In: Goodfellow M, O'Donnel AG (eds) Handbook of New Bacterial Systematics, Academic Press, London, pp. 195-250

Vestal JR, White DC (1989) Lipid analysis in microbial ecology. Bioscience 39: 535-541

Vogel H, Desaules A, Häni H. (1989) Schwermetallgehalte in den Böden der Schweiz. Bericht 40 des Nationalen Forschungsprogrammes Boden, Liebefeld-Bern

White DC (1995) Chemical ecology: possible linkage between macro- and microbial ecology. Oikos 74: 177-184

Zelles L, Bai QY, Beck T, Beese F (1992) Signature fatty acids in phospholipids and lipopolysaccharides as indicators of microbial biomass and community structure in agricultural soils. Soil Biol Biochem 24: 317-323

Zelles L, Bai QY (1993) Fractionation of fatty acids derived from soil lipids by solid phase extraction and their quantitative analysis by GC-MS. Soil Biol Biochem 25: 495-507

Appendix. PLFAs, which form the biological guilds plant roots (R), Algea (A), Protozoa (P), Fungi (F), vesicular arbuscular Mycorrhiza (VAM), Cyanobacteria (CB), and Bacteria (B) (Suzuki *et al.*1993, Ratledge and Wilkinson 1988, Vestal and White 1989, Zelles *et al.* 1992, Olsson *et al.* 1995, Pennanen *et al.* 1996); Numbers in brackets are relative retention times in relation to n19:0.

R: 18:3(0,949), 18:3(0,956), 18:3_6,9,12, 18:3_10,12,16, n24:0, n26:0, n26:0(1,534)

A: 18:2(0,882), 18:2(0,898), 18:2(0,901), 18:2(0,903), 18:2(0,932), 18:2_5,10, 18:2_6,10, 18:2_9,12, d6:0, d8:0, d9:0, d10:0, d11:0, d12:0, d13:0, d14:0, d15:0, d16:0, d17:0, d18:0, d20:0, d21:0, d22:0

P: 20:4_4,8,11,14, 20:3(1,053), 20:3(1,065), 20:3(1,066), 20:2(1,042), 20:2_5,10, 20:2_10,14, 22:2(1,186), 22:2(1,192), 22:2(1,202), n14:0

F: n23:0, 20:4_4,8,11,14, 18:2(0,882), 18:2(0,898), 18:2(0,901), 18:2(0,903), 18:2(0,932), 18:2_5,10, 18:2_6,10, 18:2_9,12, 16:1,11(1,271), n14:0, n16:0, n18:0, n10:0, n11:0, n12:0, n13:0, n15:0, n17:0, n19:0, n20:0, n21:0, n22:0, 18:1,9_(1,370), 18:1,9_(1,401), 18:1,9_(1,406), 18:1,9_(1,418), 18:1,11(1,401)cis, 18:1,11(1,406)cis, 18:1,11(1,418)cis

VAM: 20:4_4,8,11,14, 16:1,11(1,271), 18:1,11(1,401)cis, 18:1,11(1,418)cis

CB: 20:3(1,053), 20:3(1,065), 20:3(1,066), 16:3(0,725), 16:3_5,7,10, 20:3_5,8,11, 20:3_5,11,15, 22:3(1,111), 18:3(0,949), 18:3(0,956), 18:3_6,9,12, 18:3_10,12,16, 14:1,4_(1,098), 15:1,4_(1,147), 15:1,7_(1,147), 15:1,4_(1,154), 15:1,7_(1,154), 15:1,4_(1,176), 15:1,7_(1,176), 16:1,4_(1,225), 16:1,7_(1,225), 16:1,4_(1,253), 16:1,7_(1,253), 16:1,11(1,277), 17:1,4_(1,299), 17:1,9_(1,299), 17:1,11(1,299), 17:1,4_(1,307), 17:1,9_(1,307), 17:1,11(1,307), 17:1,4_(1,327), 17:1,9_(1,327), 17:1,11(1,327), 17:1,4_(1,334), 17:1,9_(1,334), 17:1,11(1,334), 18:1,4_(1,370), 18:1,4_(1,401), 18:1,4_(1,418), 18:1,13(1,418), 19:1,13(1,423), 19:1,7_(1,446), 19:1,4_(1,449), 19:1,7_(1,449), 19:1,4_(1,471), 19:1,11(1,477), 20:1,8_(1,570), 20:1,11(1,570), n14:0, n16:0, n18:0, n10:0, n11:0, n12:0, n13:0, n15:0, n17:0, n19:0, n20:0, n21:0, n22:0, i15:0, a15:0, i17:0, a17:0, i13:0, a13:0, i14:0, i16:0, i18:0

B: 16:1,11(1,271), 14:1,4_(1,098), 15:1,4_(1,147), 15:1,7_(1,147), 15:1,4_(1,154), 15:1,7_(1,154), 15:1,4_(1,176), 15:1,7_(1,176), 16:1,4_(1,225), 16:1,7_(1,225), 16:1,4_(1,253), 16:1,7_(1,253), 16:1,11(1,277), 17:1,4_(1,299), 17:1,9_(1,299), 17:1,11(1,299), 17:1,4_(1,307), 17:1,9_(1,307), 17:1,11(1,307), 17:1,4_(1,327), 17:1,9_(1,327), 17:1,11(1,327), 17:1,4_(1,334), 17:1,9_(1,334), 17:1,11(1,334), 18:1,4_(1,370), 18:1,4_(1,401), 18:1,4_(1,418), 18:1,13(1,418), 19:1,13(1,423), 19:1,7_(1,446), 19:1,4_(1,449), 19:1,7_(1,449), 19:1,4_(1,471), 19:1,11(1,477), 20:1,8_(1,570), 20:1,11(1,570), n14:0, n16:0, n18:0, n10:0, n11:0, n12:0, n13:0, n15:0, n17:0, n19:0, n20:0, n21:0, n22:0, i15:0, a15:0, i17:0, a17:0, i13:0, a13:0, i14:0, i16:0, i18:0, br9:0, p2-15:0, p6-13:0, br13:0(0,483), p3-15:0(0,599), p2-18:0, p8-15:0(0,608), br15:0(0,610), br15:0(0,613), p2-16:0(0,642), br16:0(0,646), p8-16:0(0,661), br16:0(0,663), br16:0(0,692), br16:0(0,707), p2-17:0(0,725), br17:0(0,731), br17:0(0,748), br17:0(0,750), br17:0(0,753), p2-17:0(0,773), p2-18:0(0,776), br17:0(0,782), br17:0(0,785), br17:0(0,787), br17:0(0,789), br17:0(0,791), br17:0(0,795), br17:0(0,815), p2-18:0(0,818), br18:0(0,834), br18:0(0,835), br18:0(0,836), met16:0, p2-18:0(0,859), p2-19:0(0,861), br18:0(0,871), p3-20:0, br18:0(0,893), br19:0(0,910), p2-20:0(0,943), br19:0(0,949), br19:0(0,953), br19:0(0,969), p2-20:0(0,973), br19:0(0,976), p6-20:0, br20:0(1,029), br20:0(1,031), br20:0(1,049), p4-21:0(1,063), p2-21:0(1,103), br21:0(1,107), p4-21:0(1,116), p2-21:0(1,122), p2-22:0(1,129), p2-22:0(1,130), br21:0(1,135), p4-22:0(1,140), p4-22:0(1,141), p4-22:0(1,165), p4-22:0(1,237), 16:1,9_(1,259)cis, 16:1,9_(1,261)cis, 16:1,9_(1,263)cis, 18:1,9_(1,370), 18:1,9_(1,401), 18:1,9_(1,406), 18:1,9_(1,418), 18:1,10(1,404), 14:1,5_(1,098), 14:1,9_(1,114), 15:1,5_(1,147), 15:1,9_(1,147), 15:1,5_(1,154), 15:1,9_(1,154), 15:1,5_(1,176), 15:1,9_(1,176), 15:1,9_(1,185), 16:1,5_(1,225), 16:1,9_(1,225)trans, 16:1,5_(1,253), 16:1,10(1,261), 17:1,5_(1,286), 17:1,5_(1,299), 17:1,5_(1,307), 17:1,5_(1,327), 17:1,5_(1,334), 18:1,5_(1,370), 18:1,5_(1,401), 18:1,5_(1,406), 18:1,5_(1,418), 19:1,5_(1,449), 19:1,9_(1,449), 20:1,5_(1,570), p10-17:0, p10-18:0, p10-19:0, cy19:0(0,906), cy19:0(0,923), cy19:0(0,925), cy19:0(0,963), cy19:0(0,989), cy19:0, cy19:0(0,994), cy17:0(0,824), cy17:0(0,825), cy17:0(0,829), cy16:0(0,736), cy18:0(0,874), cy18:0(0,898), cy18:0(0,907), 18:1,11(1,401)cis, 18:1,11(1,406)cis, 18:1,11(1,418)cis, p10-16:0, 18:1,12(1,418)cis, 18:1,11(1,370)trans

Impact of Fertilizers on the Humus Layer Microbial Community of Scots Pine Stands Growing Along a Gradient of Heavy Metal Pollution

Hannu Fritze[1], Taina Pennanen[1] and Pekka Vanhala[2]

[1] Finnish Forest Research Institute, P.O. Box 18, FIN-01301 Vantaa, Finland
[2] Finnish Environment Agency, P.O. Box 140, FIN-00251 Helsinki, Finland

Abstract. Forest vitality fertilization experiments were established in 1992 on a heavy metal deposition gradient in order to estimate the impact of the fertilizers on the disturbed forest ecosystem. The treatments consisted of control (C), lime (L), ground apatite (TF) and nitrogen (NL at 0.5 and 4 km; N at 8 km) applications at distances of 0.5, 4 and 8 km from a Cu-Ni smelting plant. Along the 8 km long transect towards the plant the total Cu concentration of the humus layer of the Calluna site type Scots pine (*Pinus sylvestris*) stands increased from 294 ± 40 mg kg^{-1} dry matter in the less polluted area to 7634 ± 880 mg kg^{-1} d.m. in the heavily polluted area.

The microbial community was characterized by phospholipid fatty acid (PLFA) and community-level carbon source utilization (BIOLOG; GN microtiter plates) patterns. The data were analyzed by principal component analysis (PCA). The pollution induced a gradual change in the PLFA pattern of the humus samples, indicating a change in microbial community structure. In the heavily polluted area the fertilization treatments had no effect on the microbial community structure since no differences between the treatment and control plots were observed. In the medium and less polluted study area a different PLFA composition was observed in the NL and L plots, indicating a treatment effect. However, the community structures in the limed plots of the less and medium polluted areas were not similar.

With the BIOLOG approach the PCA clustered the 36 study plots into three groups. One group consisted of all the polluted plots with no treatment-related separation. The second group consisted of plots from the medium and less polluted area including the C, TF and N treatments. The third group consisted of the plots from the medium and less polluted study area with the L or NL treatment. When using the BIOLOG approach no distinction between the medium and less polluted areas was achieved. Thus a 3.8 fold increase in the pollution level between the less and medium polluted area induced a change in the microbial community structure (PLFA pattern) without influencing the functional diversity (BIOLOG pattern). Liming treatments induced similar changes in the functional diversity of the microbial communities of the less and medium polluted areas even though they had different initial PLFA patterns.

Keywords. Apatite, Biolog, copper, heavy metals, nitrogen, lime, PLFA, substrate-use

1. Introduction

Heavy metal pollution decreases soil microbial activity and microbial biomass and induces changes in the soil microbial community structure (see Bååth 1989 for a review). This leads to a decreased mineralization rate of deposited organic material. Therefore heavy metal polluted sites in coniferous forests are characterized by an undecomposed layer of needle litter (see Fritze 1992). The toxicity of heavy metals is strongly dependent on soil texture and its physicochemical properties. The physicochemical soil properties influencing the toxicity of heavy metals include: pH, oxidation-reduction potential, inorganic anions and cations, water hardness, clay minerals, organic matter and temperature (Collins and Stotzky 1989). One way of influencing soil physicochemical status is fertilization treatment. Therefore treatments such as liming have been included in restoration experiments of sites polluted with heavy metals (Winterhalder 1995).

Forest vitality fertilization experiments were established in 1992 on a heavy metal deposition gradient in southwestern Finland in order to estimate the impact of the fertilizers on the disturbed forest ecosystem. The treatments consisted of a control, lime, ground apatite and nitrogen application with three replicates at distances of 0.5, 4 and 8 km from a Cu-Ni smelting plant (Fritze *et al.* 1996). The increase in total Cu concentration in the humus (F/H) layer of the *Calluna* site type Scots pine (*Pinus sylvestris*) stands was from ca. 300 to 8 000 mg kg^{-1} d.m. (dry matter) along the 8 km long transect towards the plant. Within this Cu pollution range, all the measured microbial biomass variables declined to 10% - 28% of the control plot values and activity assessed by respiration was lowered to 16% (Fritze *et al.* 1996).

The fertilization treatments, which increased the humus pH along the gradient at Harjavalta, increased soil respiration rates but had little effect on the microbial biomass values during the two years (Fritze *et al.* 1996). In earlier studies liming has been shown to alter the microbial community structure (Frostegård *et al.* 1993a). A change in the soil microbial community structure of heavy metal polluted sites could be the key to restoration of the soil ecosystem.

In the past decade new techniques have been introduced for studying the soil microbial community without requiring the isolation and culturing of individual species. One method is the analysis of phospholipid fatty acid (PLFA) composition in the soil by gas chromatography (Tunlid and White 1992). Different subsets of the microbial community have different PLFA patterns, and it is possible to characterize the features of the microbial community structure directly in a natural habitat (Bååth *et al.* 1992, Bååth *et al.* 1995, Pennanen *et al.* 1996). A second approach characterizing the soil microbial community was proposed by Garland and Mills (1991), who introduced the use of community-level carbon source utilization patterns for comparison of microbial communities from different habitats. They used commercially available microtiter plates containing 95 carbon substrates (BIOLOG GN; BIOLOG, Inc., Hayward, Calif.), which they directly inoculated with samples from different environmental sources. The produced BIOLOG pattern can be used as a measure to compare the functional diversity of microbial communities.

Two questions guided our work 1) which of the two techniques is the most sensitive one as to early warnings of heavy metal effects can be derived and 2) is the microbial community structure and function on the fertilized plots different from that on the heavy metal polluted control plots.

2. Materials and methods

Pollution source, study sites and fertilization experiments. The Harjavalta Cu-Ni smelting plant is located in southwestern Finland (61°19'N, 22°9'E) and has been polluting the environment with heavy metals for nearly 50 years. The airborne metal pollution is clearly reflected in the coniferous forest ecosystems: disappearance of sensitive epiphytic lichens growing on the bark of trees, practically non-existent ground vegetation and a decrease of humus microbial activity concomitant with an increase in the heavy metal concentrations in the humus layer close to the plant (Fritze *et al.* 1989). Since 1987 the metallic dust emissions have been reduced and in 1993 the emissions were as follows (in kg year^{-1}): Cu 50 000, Ni 7 000, Zn 13 000, Pb 6 000, and As 11 000.

The present study was performed in *Calluna* site type Scots pine (*Pinus sylvestris*) stands lying on a 8 km long transect starting from the point emission source. Fertilization experiments were performed in spring 1992 mostly on 30 x 30 m plots with four treatments and three replicates at distances of 0.5 (heavily polluted), 4 (medium polluted) and 8 km (less polluted) from the plant. The treatments within each distance group consisted of:

I. C = Control
II. L = 2 000 kg granulated limestone + 2 kg boron ha^{-1}
III. TF = 31 kg P + 56 kg K + 98 kg Ca + 61 kg Mg + 37 kg S + 0.8 kg Cu + 0.8 kg Zn + 1.3 kg B ha^{-1} (slow releasing test fertilizer made from ground apatite)
IV. N = Nitrogen fertilization. The nitrogen fertilization was designed on the basis of needle analyses for each Scots pine stand. The doses used were at 0.5 km: 150 kg N + 50 kg Mg + 1 500 kg limestone ha^{-1} (NL$_1$); at 4 km: 150 kg N + 30 kg P + 30 kg Mg + 1 000 kg limestone ha^{-1} (NL$_2$); at 8 km: 120 kg N ha^{-1} (N). The nitrogen fertilizer was always methylene urea-N (100 kg N ha^{-1}) and NH$_4$NO$_3$-N (the remainder of N).

Soil sampling and chemical determinations. At each plot one bulk sample, consisting of 24 separate samples (soil core diameter 7.2 cm), was taken from the entire humus layer (ca. 3 cm thick F/H-horizons) on the 24th of August 1994. The bulk soil samples were immediately sieved (4 mm mesh) and stored at 4 °C. Subsamples of the field-moist humus were heated at 105 °C for 12 h to determine the dry mass. For the chemical characterization of the humus samples see Fritze *et al.* (1996) and Table 1.

Table 1. Cation exchange capacity, base saturation, pH and carbon and nitrogen contents of the humus.

Variable	0.5 km				4 km				8 km			
	C	L	TF	NL_1	C	L	TF	NL_2	C	L	TF	N
CEC	16.5	26.7	17.1	21.4	30.0	39.2	31.0	35.9	26.5	47.9	29.9	27.9
	(2.03)	(2.11)	(2.34)	(3.43)	(0.95)	(2.25)	(0.30)	(1.34)	(0.63)	(1.61)	(0.44)	(0.92)
BS	12.4	50.5	27.2	47.6	41.1	87.0	53.8	74.7	40.4	92.6	58.4	46.7
	(2.37)	(4.53)	(3.69)	(8.91)	(1.24)	(2.13)	(1.88)	(3.20)	(0.84)	(0.37)	(0.91)	(1.38)
$pH_{(H_2O)}$	3.92	4.19	4.01	4.31	3.67	4.46	3.77	4.11	3.74	4.81	3.83	3.71
	(0.05)	(0.09)	(0.04)	(0.12)	(0.09)	(0.05)	(0.10)	(0.08)	(0.02)	(0.01)	(0.02)	(0.02)
C_{org}	322	406	336	347	499	460	453	489	452	482	479	483
	(53)	(19)	(40)	(47)	(8.0)	(9.9)	(26)	(1.8)	(8.0)	(12)	(22)	(11)
N_{tot}	10.7	13.0	12.0	11.4	13.4	11.6	11.5	13.4	11.9	12.5	12.6	12.6
	(1.2)	(1.3)	(1.1)	(1.4)	(0.7)	(0.3)	(0.3)	(0.2)	(0.2)	(0.1)	(0.5)	(0.2)

Cation exchange capacity (CEC) is given in milliequivalents $(100\ g)^{-1}$ d.m. and the base saturation (BS) is given in % of CEC, C_{org} and N_{tot} are presented in g kg^{-1} d.m. Standard error of the mean in parentheses.

PLFA analysis. Phospholipid fatty acids (PLFAs) were extracted and analyzed by a procedure described by Frostegård et al. (1993b). Briefly, 0.5 g (fresh weight) of humus was extracted with a chloroform-methanol-citrate buffer mixture (1:2:0.8) and the lipids were separated into neutral lipids, glycolipids and phospholipids on a silicic acid column. The phospholipids were subjected to mild alkaline methanolysis and the fatty acid methyl esters were separated by gas chromatography (GC, flame ionization detector) using a 50m HP-5 (phenylmethyl silicone) capillary column. Helium was used as a carrier gas. Methyl tridecanoic fatty acid 13:0 and methyl nonadecanoate fatty acid (19:0) were used as internal standards and peak areas were quantified with the help of 19:0. The fatty acid nomenclature is as follows: fatty acids are designated as the total number of carbon atoms:number of double bonds, followed by the position of the double bond from the methyl end of the molecule. Cis and trans configurations are indicated by c and t, respectively. The prefixes a and i indicate anteiso- and iso-branching, 10Me indicates a methyl group on the tenth carbon atom from the carboxyl end of the molecule, and cy refers to cyclopropane fatty acids.

BIOLOG. BIOLOG GN microtiter plates containing 95 different C-sources and a redox dye (BIOLOG Inc., Hayward, California) were used. A description of the GN substrates, which include carbohydrates, organic acids, amino acids, polymers, esters, alcohols, amides, amines, aromatics and phosphorylated and brominated compounds, was given by Garland and Mills (1991). The plates were inoculated with a 10^{-4} dilution (0.9% NaCl) made from the humus samples. The total numbers of bacteria (microscopic count) were ca. 1.8 times lower near the smelting plant (0.5 km) than in the distance groups 4 and 8 km, where they were at approximately the same level (Pennanen *et al.* 1996). No correction was made for the inoculation density. The plates were incubated at 20 °C and read after 24, 96, 146, 194 and 256 h. Absorbance values for the wells with C sources were blanked against the control well. Negative values were considered as 0 in the subsequent data analysis.

Statistics. The mol% (percentage of the total area of the GC chromatogram) of the PLFA values were log_{10} transformed before being subjected to principal component analysis (PCA). For the BIOLOG plates the percentage absorbance for each well of the plate total absorbance was calculated before subjecting the values to PC analysis.

3. Results

Several separate PCAs were performed with the PLFA data. First the whole data set was analyzed. The first two PCA axes explained 85 % of the variation and separated the study plots into 2 groups (Fig. 1a). One group included all the study plots from the heavily (0.5 km) polluted study area whereas all the other study plots of the medium (4 km) and less (8 km) polluted area formed the other group. Within the second group the plots from the medium polluted area separated from the plots of the less polluted area along the second axis. The scores of the first axis can be correlated with the humus copper concentration (see Pennanen *et al.* 1996) and therefore the first axis was related to the pollution, explaining 79 % of the data variation. Five fatty acids were mainly responsible for the separation of the study plots into the two groups. Due to pollution the relative mol percentages of the branched fatty acids br18:0, br17:0 and i17:0 increased whereas 18:2ω6 and 20:4 strongly decreased (Fig 1b). The second PCA was performed without the heavily polluted plots. This resulted in a separation of the medium polluted study plots from the less polluted ones along the first axis explaining 46 % of the variation (Fig. 2a). The PLFAs most responsible for the grouping were: the relative mol% of the PLFAs 20:5, br18:0, br17:0 and 14:0 were higher in the humus of the study plots from the medium polluted area whereas 20:4 and 16:1ω5 were at a higher level in the less polluted study plots (Fig 2b). No clear, consistent treatment-related separation of the study plots could be detected in either PCA. Only in the less polluted area did a separation of the limed plots from the controls along the second axis begin to appear (Fig. 2a).

In order to exclude the effect of pollution, the heavily, medium and less polluted areas were also investigated separately by PCA. In the heavily polluted area no real treatment related grouping of the study plots was detected (Fig. 3), which

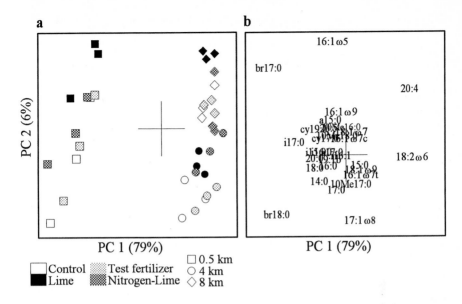

Fig. 1. a) Score plot of PCA showing the separation of all study plots along the first two principal components (PC 1 and PC 2) using PLFA data. At the distance of 8 km no lime was added to the nitrogen treatment. See Materials and Methods for the exact composition of the fertilizers. b) Loading values for the individual PLFAs. PLFAs most distant from the origin are responsible for the separation of the study plots.

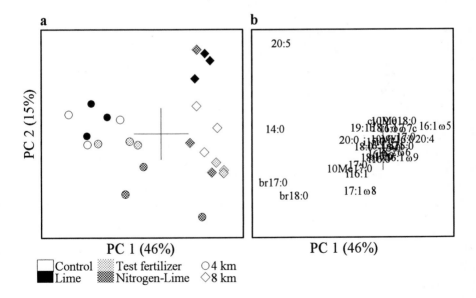

Fig. 2. PCA score (a) and loading (b) plots of the PLFAs from the medium and less polluted study areas.

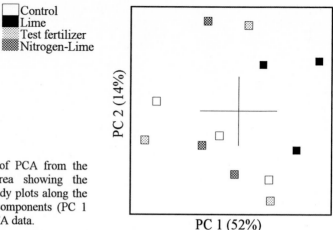

Control
Lime
Test fertilizer
Nitrogen-Lime

PC 2 (14%)

PC 1 (52%)

Fig. 3. Score plot of PCA from the heavily polluted area showing the separation of the study plots along the first two principal components (PC 1 and PC 2) using PLFA data.

could be related to the increased humus pH (Table 1). In the other two areas a separation along the first axis of the study plots having received a liming treatment (alone or in combination with nitrogen) from the corresponding controls was demonstrated (Figs. 4a and 5a). In the medium polluted area the relative mol% of the PLFAs 16:1ω5 and 20:4 increased due to the L and NL treatments whereas the fatty acids cy19:0, br18:0, i15:0, and 16:1ω7t were higher in the control plots (Fig. 4b). Two of the three study plots having received the ground apatite fertilization (TF) were in the PCA close to the control plots and did not form a distinctive

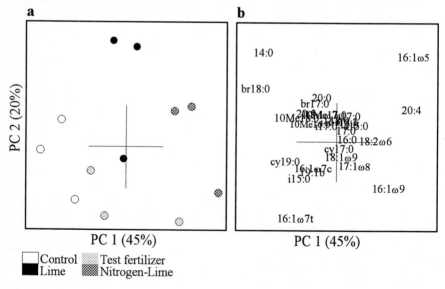

a

PC 2 (20%)

PC 1 (45%)

b

14:0   16:1ω5

br18:0

20:0
br17:0
10Me... 20:0 ...17:0   20:4
10Me... ...17:0
...18:0
16:0 18:2ω6
cy17:0
cy19:0   18:1ω9
16:1ω7c 17:1ω8
i15:0
16:1ω9

16:1ω7t

PC 1 (45%)

Control   Test fertilizer
Lime   Nitrogen-Lime

Fig. 4. PCA score (a) and loading (b) plots of the PLFAs from the medium polluted study area.

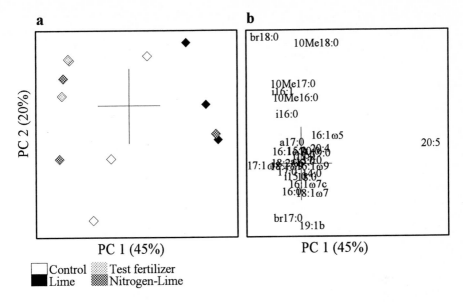

a

PC 2 (20%)

PC 1 (45%)

b

br18:0   10Me18:0

10Me17:0
i16:1
10Me16:0

i16:0

a17:0   16:1ω5
16:1ω5a20:4
17:1ω8   18:2ω6:1ω9
i15:16:00
16:1ω7c
16:08:1ω7

br17:0
19:1b

20:5

PC 1 (45%)

☐ Control ▦ Test fertilizer
■ Lime   ▩ Nitrogen-Lime

Fig. 5. PCA score (a) and loading (b) plots of the PLFAs from the less polluted study area.

group (Fig. 4a). In the less polluted area the liming plots also grouped separately from the controls (Fig. 5a) but the contrast along the first axis was formed between the TF and L fertilization treatments. The mol% of the PLFA 20:5 increased due to liming and decreased due to the apatite fertilization (TF) when compared to the controls. The fatty acid 16:1w5 increased only in the L plots, whereas 10Me18:0 increased in both the L and TF plots (Fig 5b). The PLFA i16:1 is an example of a fatty acid which increased only in the TF treatment. The nitrogen fertilized plots in the less polluted area did not form a distinct group in the PCA (Fig. 5a).

For the BIOLOG data a threshold value of 0.3 absorbance units was specified before the reaction of the substrate used was considered positive. Using this criterion the samples from the less polluted (8 km) and medium polluted (4 km) areas required 96 hours of incubation before reaching an average value of 50 positive wells. The polluted samples (0.5 km) had to be incubated for 256 h to reach the same number of substrates used. The PCA of the BIOLOG data was therefore made using the data set which combined these two different reading times, i.e. 96 h-reading for samples from 4 and 8 km distance and 256 h-reading for the samples from 0.5 km distance. The PCA extracted two axes explaining 28 % of the total variation and clustered the data into three groups. One group consisted of all the polluted plots, with no treatment-related separation. The second group consisted of plots from the medium and less polluted area including the control, ground apatite and nitrogen fertilization treatments. The third group consisted of the plots from the medium and less polluted study area having received the liming or nitrogen liming treatment. Reducing the data set of 95

76

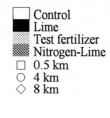

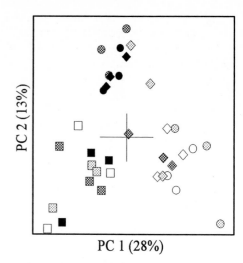

☐ Control
■ Lime
▒ Test fertilizer
▓ Nitrogen-Lime
☐ 0.5 km
○ 4 km
◇ 8 km

PC 2 (13%)

PC 1 (28%)

Fig. 6. Score plot of PCA showing the separation of all study plots along the first two principal components (PC 1 and PC 2) using BIOLOG data.

Table 2. List of BIOLOG substrates responsible for the separation of the sample plots in Fig. 6. Numbers indicate the loadings on the two principal components.

C-sources	PC1	PC2	C-sources	PC1	PC2
Carbohydrates			Amides		
N-acetyl-D-glucosamine	0.083	-0.132	Succinamic acid	-0.144	-0.166
L-Arabinose	-0.127	0.086	Alaninamide	-0.090	-0.123
Cellobiose	-0.133	-0.153			
Gentiobiose	-0.143	-0.101	Amino acids		
m-Inositol	0.197	0.061	D-Alanine	0.003	0.149
Maltose	-0.160	-0.224	L-Aspartic acid	-0.002	0.269
D-Mannose	-0.169	-0.092	L-Glutamic acid	0.337	-0.065
D-Melibiose	-0.189	-0.156	Hydroxy-L-proline	0.253	0.058
L-Rhamnose	-0.151	0.076	L-Ornithine	-0.038	0.139
D-Sorbitol	-0.031	0.121	L-Phenylalanine	0.056	-0.181
Sucrose	-0.015	-0.115	L-Pyroglutamic acid	-0.003	0.151
			L-Serine	-0.150	0.225
Esters			α-Aminobutyric acid	0.123	0.034
Methylpyruvate	0.156	-0.068			
			Aromatic chemicals		
Carboxylic acids			Urocanic acid	0.138	0.250
cis-Aconitic acid	0.146	0.170			
Citric acid	0.161	0.040	Amines		
D-galactonic acid lactone	-0.079	0.185	Phenylethylamine	-0.104	0.142
D-gluconic acid	-0.075	0.135			
Itaconic acid	-0.104	0.248	Polymers		
α-Ketobutyric acid	-0.094	-0.134	Dextrin	-0.131	-0.174
Quinic acid	0.306	-0.102			
D-Saccaric acid	0.258	-0.096			
Succinic acid	0.237	-0.104			

variables to the 35 most important ones by eliminating variables which did not significantly explain total variance (loadings < ±0.11 on PC 1 and 2) increased the percentage variance explained by the first two axes to 41 % without changing the grouping (Fig. 6). The list of substrates responsible for the separation of the sample plots along the two axes is presented in Table 2. A positive loading for an axis indicates that the substrate is used to a greater extent in samples having positive scores for that axis and correspondingly negative loadings indicate greater utilization in samples with negative scores. Thus L-glutamatic acid, quinic acid and hydroxy-1-proline are good representatives of substrates utilized in the cleaner study areas whereas the utilization of D-melobiose, D-mannose and maltose characterize the polluted study plots. The plots from the medium and less polluted study area having received the liming or nitrogen liming treatment were characterized by the utilization of L-serine or L-aspartatic acid (Table 2).

When all the heavily polluted plots were removed from the data set the PCA resulted in a separation of the plots from the medium and less polluted study area having received the liming or nitrogen liming treatment from the rest of the plots along the first axis, which explained 35% of the variation (data not shown). No pollution-related separation of the study plots was detected between the medium and less polluted areas.

4. Discussion

As well as Cu, the smelting plant at Harjavalta also emits Zn, Ni, Cd, Pb, Fe and S (Fritze *et al.* 1989). As the measured pollutants are well correlated with the distance from the plant, Cu was chosen to represent the overall pollution. Along the 8 km long transect the total Cu concentration of the humus increased from 294 ± 40 mg kg^{-1} dry matter (less polluted area) to 1110 ± 39 mg kg^{-1} d.m. (medium polluted area) and finally reached 7630 ± 880 mg kg^{-1} d.m. at the heavily polluted area (Fritze *et al.* 1996). Figure 1 shows the PCA of the PLFA data. The plots from the polluted area had high negative scores along the first axis and are therefore projected left of the origin, whereas the plots from the medium and less polluted area are projected to the right of the origin (Fig. 1). When the study plots of the heavily polluted area were removed from the PCA a distinction between the medium polluted study area and the less polluted area was evident. The pollution gradient thus induced a gradual change in the PLFA pattern of the humus samples, indicating a change in microbial community structure. A similar result was obtained by Pennanen *et al.* (1996), who sampled only the controls of this gradient, including plots at a distance of 2 km from the plant. Their results are based on almost the same changes in the relative PLFA concentrations as in our study. Therefore their discussion of the relationship between the PLFAs and heavy metal pollution is not repeated here in detail. In brief, the heavy metal pollution induced change in the microbial community structure is strongly related to the decrease of the PLFA 18:2ω6, which can be related to the presence of fungi in the sample. Since the plant vegetation and coniferous root system is damaged due to the heavy metal pollution, Pennanen *et al.* (1996) proposed that the amount of ectomycorrhizal fungi decreased along the gradient towards the smelting plant. It should be noted that the pollution-related distinction of the sample plots could also be achieved when the fungal fatty acid was removed from the data set (data not shown). This indicated that a change in the bacterial community structure also occurred. Pennanen *et al.* (1996) were able to relate the change in the bacterial community structure along this gradient to the appearance of heavy metal tolerant bacteria.

We were interested to determine whether the fertilization experiments along this pollution gradient had influenced the microbial community. The effects of the fertilization procedures on the cation exchange capacity (CEC), base saturation (BS), pH, and C_{org} and N_{tot} of the humus layer are presented in Table 1. The L and NL treatments increased the CEC, BS and pH of the humus as compared to the respective controls, whereas the treatments TF and N (N was only used at the less polluted study area) had no effect on these variables. Therefore the L and NL treatments had the potential to affect the microbial community, since increase in soil pH is correlated with decrease of heavy metals in the soil solution (Winterhalder 1995). Furthermore treatments increasing soil pH such as liming and wood ash fertilization of unpolluted coniferous humus, have been reported to change the microbial community (Bååth and Arnebrant, 1993 and 1994; Frostegård *et al.* 1993a). In the study by Frostegård *et al.* (1993a), as well as in another study in which a coniferous forest area received alkaline deposition from an iron and steel works (Bååth *et al.* 1992), most of the changes in the microbial community structure could be related to the increase in the PLFAs 10Me18:0 and 16:1ω5. Therefore these two PLFAs will be discussed here. The PLFA 10Me18:0 indicates the presence of actinomyces in the sample (Federle 1986). The PLFA 16:1ω5 is found in arbuscular mycorrhizal fungi (Olsson *et al.* 1995) and in some bacteria (Intriago 1992, Nichols *et al.* 1986).

In this study, as well as in the study by Pennanen *et al.* (1996), 10Me18:0 did not react to the heavy metal pollution. However, in the less polluted area the relative abundance of 10Me18:0 increased due to liming. There was no increase of this fatty acid in the limed plots of the medium or heavily polluted area when compared to the corresponding controls. This indicated that only in the less polluted study area did a change towards an actinomycetes richer microbial community occur due to liming. However, the increase in this study was only from 0.95 mol% to 1.14 mol%, whereas in the studies by Frostegård *et al.* (1993a) and Bååth *et al.* (1992) this PLFA increased twofold due to the liming. At least three reasons for the smaller increase in this study can be proposed: i) the amount of lime applied in this study (2 000 kg) was lower than in the study by Frostegård *et al.* (1993a) (3 800 kg) or Bååth *et al.* (1992) (continuous lime deposition), ii) the influence time of the lime applied was shorter (2 years) in this study compared to at least 4 years in the study by Frostegård *et al.* (1993a) and over 20 years in the study by Bååth *et al.* (1992), iii) the effect of heavy metal pollution: comparing the studies from Frostegård *et al.* (1993a) and Bååth *et al.* (1992) only 5 PLFAs out of 27 reacted differently due to liming whereas in this study half of the PLFAs reacted opposite to the PLFAs in the study of Bååth *et al.*

(1992). This might indicate that the microbial community structure in the limed plots of this study are still more governed by the metal pollution than by the liming.

The PLFA 16:1ω5 in this study was also not related to the pollution. This fatty acid increased in all limed humus samples throughout the pollution gradient, in good agreement with the earlier liming studies (Bååth *et al.* 1992, Frostegård *et al.* 1993a). Only in the heavily polluted area did this fatty acid not increase due to the NL treatment, whereas in the medium polluted area the humus of the NL treatments had the highest relative mol% of the PLFA 16:1ω5. As in the case of the 10Me18:0, the 16:1ω5 also did not increase to more than 1.4 fold the control level. This is much lower than in the studies by Frostegård *et al.* (1993a) and Bååth *et al.* (1992). The increase of this PLFA in the liming treatments along the gradient is an indication that a change in the microbial community structure occurred. In the heavily polluted area the fertilization treatments apparently had no overall effect on the microbial community structure, since no clear separation of the treatment plots from the control plots was observed (Fig. 3). In the medium and less polluted study area a different PLFA pattern was observed in the NL and L plots, indicating a different microbial community due to treatment (Figs. 4 and 5). However, the community structures in the limed plots of the less and medium polluted areas were not alike (Fig. 2) and this could be exemplified with the different reaction of 10Me18:0 in the limed plots in these two areas. This was probably due to the fact that the microbial communities in the limed plots had developed from different communities due to the heavy metal pollution.

Direct incubation of environmental samples in BIOLOG plates produces patterns of potential C source utilization for microbial communities. These physiological profiles have been used to classify and characterize microbial communities (Garland and Mills 1991, Bossio and Scow 1995). We used this approach to measure the functional diversity of the microbial communities along the fertilized pollution gradient. The PCA clustered the 36 study plots into three groups (Fig. 6). One group consisted of all the polluted plots with no treatment-related separation. The second group consisted of plots from the medium and less polluted areas including the control, ground apatite and nitrogen fertilization treatments. The third group consisted of the plots from the medium and less polluted study areas having received the liming or nitrogen liming treatment. When using the BIOLOG approach no separation of the medium and less polluted area was achieved. Thus a 3.8 fold increase in the pollution (Cu) level between the less and medium polluted areas induced a change in the microbial community structure (PLFA pattern) without influencing the functional diversity (BIOLOG pattern). Liming treatments induced similar changes in the function of the microbial communities of the less and medium polluted areas despite the difference in community structures. These differences in the ability of the PLFA and the BIOLOG approaches to characterize the microbial communities in the less and medium polluted study areas point towards the fact that BIOLOG should not be used by itself as a tool to measure changes in community structure as it has been discussed in the literature (Bossio and Scow 1995): the same BIOLOG pattern can be produced by different PLFA patterns.

The 26 fold increase in the pollution level changed the microbial community structure and the physiological profiling. In the heavily polluted area less C-sources were used compared to the medium and less polluted areas. Furthermore the soils of the heavily polluted area were characterized by a lower rate of substrate utilization together with a longer lag time before the substrates were used. This behavior is typical for the microbial communities of heavy metal polluted soils (Nordgren *et al.* 1986, Nordgren *et al.* 1988, Burkhardt *et al.* 1994).

Garland (1996) discussed some important points when using the BIOLOG approach for characterizing the functional diversity of microbial communities: Firstly when the bacterial inoculum ranged between 10^3 and 10^7 acridine orange stainable (AO) cells, most of the variation in the overall rate of color development of the BIOLOG plates could be explained by inoculation density. However, repeated monitoring can minimize the variation in the degree of color development among samples by allowing for selection of a given reference point in color development. Thus an average well color development (AWCD) of 0.75 abs. units, a midpoint in overall color development at which about 60% of the wells had become colored, was used as the reference point. This approach eliminated the time-consuming step of checking and adjusting initial inoculum densities (Garland 1996). In our study the difference in AO stainable inoculation density of the control plots between the heavily polluted and less polluted area was twofold. Therefore in our evaluation of the results we used those plate readings in which ca. 50% of the wells were positive. Secondly, Haack *et al.* (1995) and Garland (1996) showed that the C sources responsible for the differences between samples varied with the reading time of the plates. This was also noted in the present study when the PCA was performed for all plots with the data collected 146 hours after inoculation (data not shown). The same separation could be achieved for the sample plots as shown in Fig. 6, although different C sources than those listed in Table 2 were responsible.

Some studies (e.g. Insam *et al.* 1996a, Wünsche *et al.* 1995, Pfender *et al.* 1996) normalize the BIOLOG absorbance data before PCA. One recommended and tested procedure of normalization was performed by dividing the values for individual wells by the AWCD for the sample (Garland and Mills 1991). This corrects for differences in inoculation densities (Garland 1996). In this study we used a different approach for the normalization of the BIOLOG data in order to make it comparable to the normalization of the PLFA data. We used the percentage absorbance of the plate total absorbance for each well. This is comparable to the normalization procedure suggested by Garland and Mills (1991). However, BIOLOG studies using GN plates cannot be directly compared with regard to which C-sources are responsible for PCA grouping, since all studies have hitherto used different reading times of their plates on normalized or unnormalized data.

The use of the metabolic quotient qCO_2 (respiration/biomass) of the microflora as an indicator of stress has lately been discussed in the literature (Brookes 1995, Wardle and Ghani 1995, Insam *et al.* 1996b). Generally it is believed that stress causes the soil microbial biomass to direct more energy from growth into maintenance so that an increased proportion of C taken up by the biomass will be respired as CO_2. This would be seen as an increased qCO_2. This is not necessarily the only interpretation of a changed qCO_2. As proposed by Insam *et al.* (1996b) a change of the qCO_2 may indicate (i) changes in the substrates utilized by an unchanged community, (ii) a change in community composition, (iii) changes in both.

Fritze *et al.* (1996) reported the qCO_2 and Pennanen *et al.* (1996) the bacterial growth rate to decrease in the control plots along the gradient towards the smelting plant. Since the qCO_2 decreases with increasing stress the maintenance energy hypothesis cannot be the only explanation even though the growth rate decreased. The decrease of the qCO_2 between the medium and heavily polluted control study plots, respectively, can be explained by a changed community structure (Pennanen *et al.* 1996; this study) and a changed substrate utilization pattern. However, the decrease in the qCO_2 from the less polluted study area towards the medium polluted area can be explained only by the altered community structure, since the patterns of substrates utilized were similar between these two areas.

Finally it can be stated that of the two microbial community characterizing techniques used in this study, PLFA and BIOLOG, the PLFA approach seems to be the better choice for detecting differences between controls and soil samples suffering from low level heavy metal pollution. Furthermore, fertilization of the heavy metal polluted forest gradient with lime or nitrogen combined with lime induced a change in the microbial community structure and its functional diversity only when the concentration of the pollution indicator Cu in the humus was below 1 100 mg kg^{-1} d.m. In the more polluted humus close to the smelting plant no change in the microbial community structure or its physiological diversity could be detected two years after the treatments.

Acknowledgements. This work was supported by a grant from the Academy of Finland to H.F. and Michael Bailey revised the English in this paper.

5. References

Bååth E (1989) Effects of heavy metals in soil on microbial processes and populations (a review). Water Air Soil Poll 47:335-379

Bååth E, Frostegård Å, Fritze H (1992) Soil bacterial biomass, activity, phospholipid fatty acid pattern, and pH tolerance in an area polluted with alkaline dust deposition. Appl Env Microbiol 58:4026-4031

Bååth E, Arnebrant K (1993) Microfungi in coniferous forest soils treated with lime or wood ash. Biol Fertil Soils 15:91-95

Bååth E, Arnebrant K (1994) Growth rate and response of bacterial communities to pH in limed and ash treated forest soils. Soil Biol Biochem 26:995-1001

82

Bååth E, Frostegård Å, Pennanen T, Fritze H (1995) Microbial community structure and pH response in relation to soil organic matter quality in wood-ash fertilized, clear-cut or burned coniferous forest soils. Soil Biol Biochem 27:229-240

Bossio DA, Scow KM (1995) Impact of carbon and flooding on the metabolic diversity of microbial communities in soils. Appl Env Microbiol 61:4043-4050

Burkhardt C, Insam H, Hutchinson TC and Reber HH (1994) Impact of heavy metals on the degradative capabilities of soil bacterial communities. Biol Fertil Soils 16:154-156

Brookes PC (1995) The use of microbial parameters in monitoring soil pollution by heavy metals. Biol Fertil Soils 19:269-279

Collins YE, Stotzky G (1989) Factors affecting the toxicity of heavy metals to microbes. In Beveridge TJ and Doyle RJ (eds) Metal Ions and Bacteria. John Wiley & Sons, New York, 31 -90

Federle TW (1986) Microbial distribution in soil - new techniques. p. 493 -498. In F Megusar and M Gantar (eds), Perspectives in Microbial Ecology. Slovene Society for Microbiology, Ljubljana, Slovenia

Fritze H, Niini S, Mikkola K, Mäkinen A (1989) Soil microbial effects of a Cu-Ni smelter in southwestern Finland. Biol Fertil Soils 8:87-94

Fritze H (1992) Effects of environmental pollution on forest soil microflora - a review. Silva Fenn 26:37-48

Fritze H, Vanhala P, Pietikäinen J, Mälkönen E (1996) Vitality fertilization of Scots pine stands growing along a gradient of heavy metal pollution: short-term effects on microbial biomass and respiration rate of the humus layer. Fresenius J Anal Chem 354:750-755

Frostegård Å, Bååth E, Tunlid A (1993a) Shifts in the structure of soil microbial communities in limed forests as revealed by phospholipid fatty acid analysis. Soil Biol Biochem 25:723-730

Frostegård Å, Tunlid A, Bååth E (1993b) Phospholipid fatty acid composition, biomass, and activity of microbial communities from two soil types experimentally exposed to different heavy metals. Appl Env Microbiol 59:3605-3617

Garland JL (1996) Analytical approaches to the characterization of samples of microbial communities using patterns of potential C source utilization. Soil Biol Biochem 28:213-221

Garland JL, Mills AL (1991) Classification and characterization of heterotrophic microbial communities on the basis of patterns of community-level sole-carbon-source utilization. Appl Env Microbiol 57:2351-2359

Haack SK, Garchow H, Klug MJ, Forney LJ (1995) Analysis of factors affecting the accuracy, reproducibility,and interpretation of microbial community carbon source utilization patterns. Appl Env Microbiol 61:1458-1468

Insam H, Amor K, Renner M, Crepaz C (1996a) Changes in functional abilities of the microbial community during composting of manure. Microb Ecol 31:77-87

Insam H, Hutchinson TC, Reber HH (1996b) Effects of heavy metal stress on the metabolic quotient of the soil microflora. Soil Biol Biochem 28:691-694

Intriago P (1992) The regulation of fatty acid biosynthesis in some estuarine strains of *Flexibacter*. J Gen Microbiol 138:109-114

Nichols PD, Stulp BK, Jones JG, White DC (1986) Comparison of fatty acid content and DNA homology of the filamentous gliding bacteria *Vitreoscilla*, *Flexibacter*, and *Filibacter*. Arch Microbiol 146:1-6

Nordgren A, Kauri T, Bååth E, Söderström B (1986) Soil microbial activity, mycelial lengths and physiological groups of bacteria in a heavy metal polluted area. Env Poll (Series A) 41:89-100

Nordgren A, Bååth E, Söderström B (1988) Evaluation of soil respiration characteristics to assess heavy metal effects on soil microorganisms using glutamic acid as a substrate. Soil Biol Biochem 20:949-954

Olsson PA, Bååth E, Jakobsen I, Söderström B (1995) The use of phospholipid and neutral lipid fatty acids to estimate biomass of arbuscular mycorrhizal fungi in soil. Mycol Res 99:623-629

Pennanen T, Frostegård Å, Fritze H, Bååth E (1996) Phospholipid fatty acid composition and heavy metal tolerance of soil microbial communities along two heavy metal-polluted gradients in coniferous forests. Appl Env Microbiol 62:420-428

Pfender WF, Fieland VP, Ganio LM, Seidler RJ (1996) Microbial community structure and activity in wheat straw after inoculation with biological control organisms. Appl Soil Ecol 3:69-78

Tunlid A, White DC (1992) Biochemical analysis of biomass, community structure, nutritional status, and metabolic activity of microbial communities in soil. Soil Biochem 7:229-262

Wardle D, Ghani A (1995) A critique of the microbial metabolic quotient (qCO2) as a bioindicator of disturbance and ecosystem development. Soil Biol Biochem 27:1601-1610

Winterhalder K (1995) Dynamics of plant communities and soils in revegetated ecosystems: a Sudbury case study. In: Gunn JM (ed), Restoration and recovery of an industrial region. Springer-Verlag, New York, pp173-182

Wünsche L, Brüggemann L, Babel W (1995) Determination of substrate utilization patterns of soil microbial communities: an approach to assess population changes after hydrocarbon pollution. FEMS Micr Ecol 17:295-306

Disturbances and their Influence on Substrate Utilization Patterns in Soil Microbial Communities

Michael V. Gorlenko, T.N. Majorova and P.A. Kozhevin

Moscow State University, Soil Science Faculty, Soil Biology Department
Vorobievy Gory, Moscow, 119899, RUSSIA

Abstract. Multisubstrate testing methods allow to characterize soil microbial communities under different disturbance conditions and compare their functional profiles with properties and criteria of non-disturbed communities. Disturbance factors used (added glucose, added detergent powder, inoculation with an alien microbial consortium and *Azospirillum brasilense* populations) turned out to have a strong and specific influence on the substrate utilization patterns in the soils studied. The differences of functional profiles caused by presence of pollutant appear to be much stronger than the original differences between the zonal soil types. It was established that all soil microbial communities under study had similar cyclic nature of successional changes irrespective of soil type and the way of succession initiation.

Keywords: BIOLOG, disturbance, microbial ecology, soil pollution, substrate, succession, utilization

1. Introduction

Dynamic behavior of perturbed communities is a branch of general ecology closely related to the study of natural and artificial disturbances in microbial habitats.

A structural-functional approach based on the concept "community as a superorganism" is a promising one for examining such dynamics. The new functional profiling methodology, using substrate utilization spectra analysis, appears to be an excellent tool for investigations of microbial community within the structural-functional framework. Good results were obtained with the BIOLOG® system and our multisubstrate testing technique (Garland and Mills, 1991; Gorlenko and Kozhevin, 1994; Winding, 1994) for monitoring of soil microbial communities under disturbed and undisturbed conditions. Bossio and Scow (1994) demonstrated the advantages of this approach.

The objective of this work was to study responses of different soil microbial communities under subject to different disturbances using the structural-functional approach. A significant aspect of our work was to establish common properties and criteria of *non-disturbed* communities. Our experiment provides functional profiling of microbial communities using *multisubstrate testing* technique and structural description by direct epifluorescent microscopy with AO and FDA staining. Combination of these two methods allowed us to describe structural-functional dynamics of the soil microbial system following several types of disturbances (including succession changes).

2. Materials and Methods

The new *multisubstrate testing* method (based on tetrazolium redox technique) was used as a functional assay. Our method uses 47 carbon sources (Table 1) instead of the 95 in the BIOLOG® system, and appeared to be sufficient for good results. The selected set of substrates was composed to achieve maximum discrimination of soil samples according to discriminant analysis. The significant difference of our method from that of BIOLOG® was that *no peptone* was added to the wells. We suggest that excluding peptone eliminates accessory and non-specific microbial strains that may utilize the peptone and helps to avoid reutilization of primary microbial biomass.

Table 1. Sole carbon sources for Multisubstrate Testing method

Carbohydrates	Carboxylicacids	Amino acids	Alcohols	Polymers	Amides
m-Inositol	Acetic	Norleucine	Glycerol	Starch	Carbamide
L-Arabinose	Aspartic	L-Cysteine		Dextran500	Acetamide
L-Rhamnose	Citric	Histidine		Tween 80	
Dulcite	Succinic	Norvaline			
D-Sorbitol	Maleic	Threonine			
α-Lactose	Propionic	Alanine			
D-Mannitol	Octanic	Asparagine			*Nucleosides*
D-Maltose	Valeric	Vvaline			Thymidine
D-Glucose	Malonic	Serine			
Sucrose	D,L-Lactic	Phenylalanine			
Xylose		Lysine			
Pullulan		Glutamic acid			*Amines*
D-Fructose		Arginine			Creatinin
Raffinose		Aminobutyric acid			
		Aminopropionic acid			

Air-dry samples of soddy-podzolic (Moscow region) and chernozem (Rostov region) soils were used in our experiments. Microbial succession was started by addition of water or an aqueous solution of glucose or detergent (Gorlenko *et al.*, 1996) at a concentration of 0.05 g g⁻¹ soil. Three different moisture levels (5, 13 and 25 mass percent) were used in several experimental series. The samples were incubated at 25°C. For the substrate utilization assay the soil suspension (1:50) was sonicated for 1 min at 22kHz and 0.1 A and centrifuged for sedimentation of the soil particles. Four replicates were used for each sample.

Microtiter plates containing the C-sources, mineral salts, tetrazolium violet and the tested soil suspension were incubated at 25°C. Color development was measured after 72 h. of incubation using a microplate scanner. Results were transformed to avoid the influence of absolute color values related to the total inoculum size (Garland, Mills,1991).

The data obtained were processed by PC using EXCEL 7.0, STATEX statistical package, and our own software. All data were statistically tested at P=0.05 .

The following disturbance factors were used : (1) added glucose,(2) added detergent powder (imitation of pollution), (3) inoculation with an alien microbial consortium (extracted from fresh chernozem soil) and populations (*Azospirillum brasilense*, Sp7 (ATCC 29145), also termed UB1, the wild type strain). *A.brasilense* was grown on LD solid medium (1% tryptone, 0.5% yeast extract, 0.25% NaCl) and potato medium. Preparation of the microbial complex was performed without growing it on a medium. There were three cycles of extracting cells from the soil using ultrasonic treatment with subsequent centrifugation (Kozhevin, 1989).

Direct fluorescence microscopy of glass slides and soil suspensions with Acridine Orange (AO) and Fluoresceine Diacetate (FDA) staining and phase contrast microscopy were performed for community structure analysis.

3. Results and Discussion

First significant result was that all of the disturbance factors described above had a strong and specific influence on the substrate utilization patterns of the soils investigated. The results of principal component analysis of experimental data are shown in Figs.1-5. The chosen systems of coordinates allowed a clear differentiation of all samples.

Differences caused by presence of pollutant appeared to be much stronger than original differences between the zonal soil types (Fig. 1). Thus, the functional profiles of polluted soils became quite similar.

It was also demonstrated that, in general, influence of any significant perturbation factor made the functional properties of tested microbial communities similar. For example, in case of inoculation with an alien microbial community (Fig.2) substrate utilization patterns shift towards the introduced one, thereby, confirming survival of the introduced alien population in a new habitat. Moreover, the experiment showed that soddy-podzolic soils and rhizosphere that have very specific microbial communities (Fig.2) became similar to high-fertility chernozem soils. It confirms our prerequisites for use of such bacterial complexes as bacterial fertilizers (Kozhevin and Korchmaru, 1995). We suggest that this method of population tracing is a promising tool in the field of bacterial fertilizers.

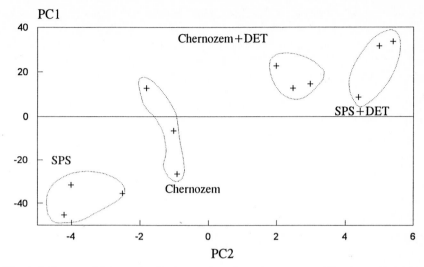

Fig. 1. Influence of detergent pollution on the sole-carbon-source utilization patterns of microbe communities of chernozem and soddy-podzolic soil (SPS). DET=detergent

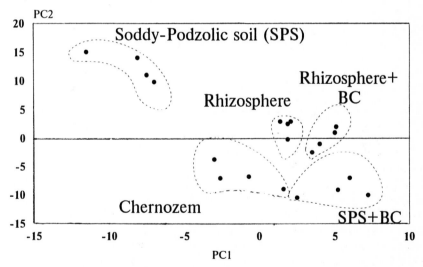

Fig. 2. Influence of introduction of fresh chernozem microbe complex (BC) into soddy-podzolic soil (SPS) and rhizosphere

Inoculation of soddy-podzolic soil with *A. brasilense* population influenced on the functional structure of microbial communities in a similar way. Multisubstrate testing allowed to visualize this population on the different stages of succession (Fig.3). Principal component analysis showed that *A. brasilense* introduction at the level $3*10^8$ cfu g^{-1} leads to significant modification of soil microbial complex characteristics. Microbial complex of inoculated soil turned out to be different from the control (Fig.3) in spite of considerable reduction (10^5 cfu g^{-1}) of *A. brasilense* in the end of the experiment (30 days). Samples of the inoculated soil fell between the projections of control soil and inoculate characteristics. The observed differences in the functional structure of microbial

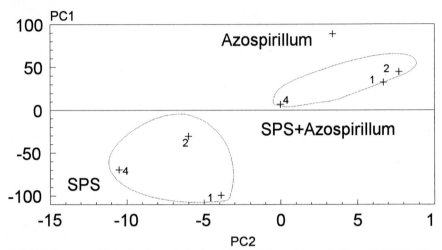

Fig.3. Influence of introduction of *A. brasilense* into soddy-podzolic soil (SPS). The figures near the points indicate the week of succession.

communities in control and inoculated soils help us to understand the fate of the inoculum. The key differences among inoculated microbial complexes were similar for all stages of succession. Table 2 shows shifts in sole-carbon-source-utilization patterns based on the image recognition analysis that define the characteristics of each type of the disturbance studied.

Our experiments make it possible to speculate on determination of some common properties of microbial communities from undisturbed and disturbed soil (Table 3). It is also useful to estimate the stability of microbial systems according to the dispersion (σ^2) of a whole substrate utilization data vector. Smooth spectra with low dispersion value characterize a stable mature community, and, vice versa, sharp spectra with high dispersion are common to 'young', disturbed environments. Polar diagrams (Fig.4) show the dynamics of 'dimension of functional capabilities' during glucose - initiated succession in soddy-podzolic soil. The maximum response was observed on day 3 when the data appeared tightly focused with maximum dispersion, followed by a period when the dimension spreading increased and dispersion decreased by day 13, reaching the condition similar to that at of day 0.

Detailed dynamics of functional profiles of communities after substrate addition were established. In spite of the different nature of successions studied (glucose, detergent or pure water initialization, different soil types) the similarity in functional dynamics was pointed out.

The cyclic nature of succession changes was visualized by factor analysis of the substrate utilization data (Fig.5). Here, lactose and glycerol have maximum factor loadings on the first two principal components. All successions described above followed the classical cyclic model forming a cycle in glycerol-lactose coordinates as shown on Fig. 7. In general, their initial stages are characterized by high polymer consumption (Fig. 6) and low potential functional capabilities (Fig.4). Later the spectrum drifts toward increasing consumption of low molecular weight substances. Stability and number of consumed substrates increased (Fig. 4).

Table 2. Disturbance criteria of soil microbial community

Factor Factor	Disturbed community criteria (substrate uptake average by AWCD)
Detergent pollution	valerianate > 0.1 and starch < 0.7
Introduction of chernozem complex	aminopropionate <1.5
Introduction A. brasilense	arabinose > 0.9 or rhamnose < 0.7

Table 3. Some key substrate utilization properties of undisturbed versus disturbed microbial communities

Community	Acetate	Lactose	Thymidine	Sorbitol	Propionat	Arabinose	Glycerol
Undisturbed	yes	high	low	high	yes	high	high
Disturbed	no	low	high	low	no	low	low

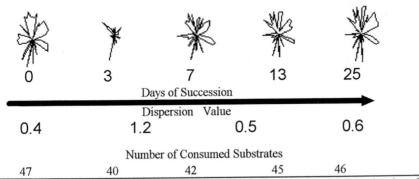

Fig.4. Visualized functional dynamic in soddy-podzolic soil during succession (the size of each ray in the polar diagrams is proportional to one substrate utilization value).

The cyclic dynamics observed in the functional assays may be explained by the analysis of community structure. Abundance curves for bacteria and microfungi during the succession events based on direct microscopy are shown on Fig 8. The early stages of succession are characterized by predominance of microfungi. They play a leading role of r-strategy organisms in primal substrate utilization. Further succession flow is most likely to lead to reutilization of accumulated fungal biomass. Bacterial biomass increased as fungal significance declined.

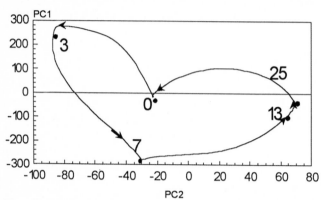

Fig. 5. Succession cycle in functional spectrum of soil microbial community. Figures mean days of succession.

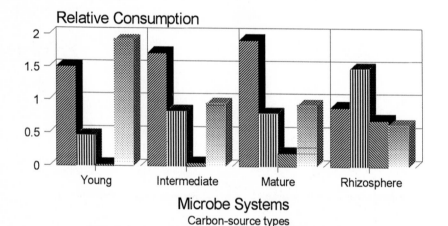

Carbon-source types

▨Carbohydrates ⁍Amino-acids ▧Organic Acids ▩Polymers

Fig. 6. Substrate consumption in microbial systems of different age and in the rhizosphere.

Based on traditional quantitative criteria, it is impossible to adequately compare the relative bacterial and fungal significance due to differences in life forms. We used a new approach that employs surface area comparison because of the important role surface membrane plays in maintaining nutrition flow. In our calculations we approximated the fungal hypha as a 5 micron diameter tube and bacterial cell - as a 0.5 micron diameter sphere.

Significance of bacteria in the community was expressed as the ratio of bacterial surface area to fungal surface area. The significance diagram (Fig.9) confirms our observation of increasing significance of bacteria during succession

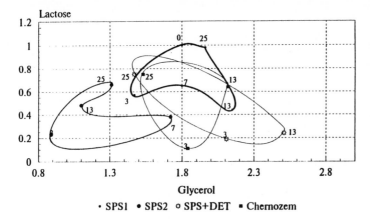

· SPS1   • SPS2   ✿ SPS+DET   ▪ Chernozem

Fig. 7. Succession trajectories in selected communities. Glycerol-lactose coordinates. SPS1-soddy-podzolic, glucose initialization , SPS2- soddy-podzolic, water initialization, SPS1-soddy-podzolic, detergent initialization, Chernozem - chernozem soil, glucose initialization.

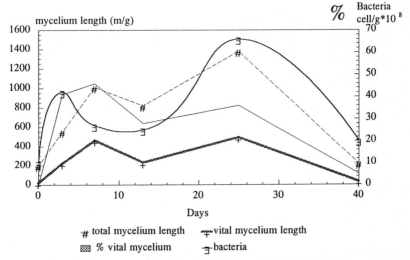

⌗ total mycelium length   ⋎ vital mycelium length
▨ % vital mycelium   ⊐ bacteria

Fig. 8. Dynamics of bacteria and microfungi abundance during microbial succession in soddy-podzolic soil.

(Fig.8). The effect of soil moisture on significance of bacteria supports the adequacy of the surface area approach (it is well known that the fungi are preferred the dry and aerobic habitats). The oscillating nature of the dynamics described suggest occurrence of cyclic processes in this system. It is known from general ecology that dynamic of this kind imply the presence of the some delay phase. The cause of this delay may be the following: the presence of significant capillary forces inside the hyphal fragments making the arbitrary leak of cytoplasm unlikely. We may consider the dead hyphae as 'canned food' inside the polymer capsule that can be slowly degradable by mostly r-strategic fungal community. Thus, further propagation of this group is temporary restricted by the absence of rich substrate. The subsequent lysis of polymer capsules by K-strategic organisms (primary by bacteria and actinomycetes) leads to the next resource enrichment that can allow new activation r-strategic organisms.

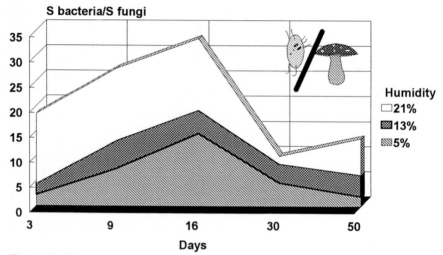

Fig. 9. Significance of bacteria in microbial succession (surface comparison approach).

4. Acknowledgements

In conclusion, we would like to thank S. Korchmaru and S. Chernets for help in our research.

5. References

Bossio DA, Scow KM (1995) Impact of carbon and flooding on the metabolic diversity of microbial communities in soil. Appl Environ Microbiol 61:4043-4050

Garland JL, Mills AL (1991) Classification and characterization of heterotrophic microbial communities on the basis of a patterns of community level sole-carbon-source utilization. Appl Environ Microbiol 57:2351-2359

Garland JL, Mills AL (1994) A community-level physiological approach for studying microbial communities. In: Ritz K, Dighton J, Giller KE (eds) Beyond the Biomass: compositional and functional analysis of soil microbial communities, Vol. , John Wiley and Sons Ltd., Chichester, UK, pp 77-83

Gorlenko MV, Kozhevin PA (1994) Differentiation of soil microbial communities by multisubstrate testing. Microbiology 63:289-293.

Gorlenko MV, Rabinovich NL, Gradova NB, Kozhevin PA, (1996) Identification of soil polluted by detergents on the basis of functional reaction of the soil microbial complex. Vestnik. Moskovskogo Universiteta, Ser.17, Pochvovedenie 1:64-69

Kozhevin PA (1989) Microbial Population in Nature (in Russian), Moscow, MSU Press.

Kozhevin PA, Korchmaru SS (1995) On theoretical substantiation of use of the use of microbial fertilizers. Moscow University Soil Sci Bull 50:45-52

Winding A. (1994) Fingerprinting bacterial soil communities using Biolog microtitre plates. In: Ritz K, Dighton J, Giller KE (eds) Beyond the Biomass: compositional and functional analysis of soil microbial communities, Vol. , John Wiley and Sons Ltd., Chichester, UK, pp 85-94

Evaluation of Remediation by Community-Level Physiological Profiles

R. Michael Lehman[1], Seán P. O'Connell[1], Jay L. Garland and Frederick S. Colwell [1]

[1] Biotechnologies Department, P.O. Box 1625, Idaho National Engineering Laboratory, Idaho Falls, ID 83404-2203, USA
[2] Dynamac Corporation, Mail Code DYN-3, Kennedy Space Center, Fla. 32899, USA

Abstract. Community-level physiological profiles (CLPP) of microbial communities based on substrate utilization have proven useful for exploring a variety of aspects of microbial ecology and may have application in the field of environmental remediation and restoration. Development of CLPP for monitoring environmental restoration is illustrated with three experimental case studies. Bioremediation of diesel fuel contaminated soils was monitored over a one year period using CLPP. CLPP of soils from contaminated and control sites showed an effect due to the presence of total petroleum hydocarbons (TPH). Treatment during the first year of the study was not effective in reducing TPH and no corresponding effect was seen in substrate profiles although a seasonal effect was noted (second year data is pending). A significant effect due to sample holding time of soils prior to analysis was seen on the CLPP of soil microbes from both contaminated and uncontaminated sites. The second field site is a fractured basalt aquifer in eastern Idaho where a contaminant plume of chlorinated solvents, sewage and radionuclides exists. CLPP on groundwater from wells with similar lithology of the screened interval reflected the presence of TCE and other pollutants with heavily contaminated wells exhibiting similar patterns that differed from uncontaminated wells. At a third site in central Idaho, aquatic bacterioplankton and periphyton communities were profiled along a stream flowpath that varied in metal concentrations due to mine drainage input. CLPP from reference sites upstream of the metal input differed from those immediately downstream of the mine drainage. Further downstream from the point source where metal concentrations declined, community profiles began to resemble the reference sites with periphyton communities recovering faster than the bacterioplankton. CLPP of indigenous microbial communities may be useful in risk assessment of contaminated sites and monitoring subsequent remedial activities.

Keywords. Ecological risk assessment, metabolic profiling, community, microbial, remediation

1. Introduction

Substrate utilization patterns derived from inoculating whole microbial communities into Biolog GN microtiter plates have been demonstrated as a rapid method of obtaining a "fingerprint" for mixed communities of heterotrophic microorganisms (Garland and Mills, 1991). Community substrate utilization patterns or community-level physiological profiles (CLPP) are a sensitive and reproducible method for discriminating between microbial communities from a variety of environments, despite the original intent of the microplates for isolate testing and the resulting selection of carbon sources (Garland and Mills, 1991; Garland and Mills, 1994; Winding, 1994; Ellis *et al.*, 1995; Haack *et al.*, 1995; Lehman *et al.*, 1995; Garland, 1996b; Insam *et al.*, 1996).

Rapid and economical methods of microbial community profiling (e.g., CLPP) have a number of potential applications in environmental assessment. Microorganisms possess inherent advantages as indicator species due to their numbers, diversity, ubiquity, and cornerstone role in global biogeochemical cycling. Information on microbial response to contaminants is desired in ecological risk assessments (ERA) to predict possible impacts for release of chemicals to the environment or measure effects of prior releases (EPA, 1989). The microbial tests proposed for ERA such as non-indigenous bioassays, i.e., Microtox, or general measures such as C mineralization are inadequate for quantifying *in situ* risks (e.g., Joern and Hoagland, 1996). Prioritization of disturbed sites for remedial action, delineation of contamination and confirmation of efffective remediation require rapid, economical and reproducible assays that reflect the ecological and human health risk of the pollutant.

Chemical criteria for groundwater and aquatic environments have been developed from laboratory bioassays to avoid adverse human health effects and reduce environmental impact although chemical residuals may misrepresent contaminant bioavailability [risk] and underestimate chemical interactive effects. Proposed adjustments to surface water criteria to account for site-specific differences in water chemistry as it affects contaminant bioavailability may allow more accurate risk assessments for single chemicals (U.S. EPA, 1994). While improvements can be made to existing standards for evaluating *in situ* impact of contaminants in water, no accepted endpoints exist for remediation of contaminated terrestrial systems and restoration of ecosystem function. CLPP methodology may provide rapid and reproducible microbial profiles for a variety of environmental assessment purposes related to remediation of pollutants and ecosystem restoration, particularly in terrestrial systems.

Samples from three field sites (terrestrial, groundwater and aquatic) that have received pollutants were examined with CLPP to determine if this assay can differentiate the presence of contaminants from variation due to natural environmental factors. The potential for CLPP as a measure of the effectiveness of environmental remediation activities is explored at the three sites which are presented as case studies. General issues with application of CLPP for environmental assessment are considered.

2. Methods

2.1 CLPP

Surface and groundwaters were inoculated directly into Biolog GN microtiter plates within 1 d of sample collection. Soils were extracted as per Lehman *et al.* (1995) and the supernatant used for inoculum. All samples were placed at 4 °C between collection and analysis. Color development in the microplate wells was measured every 2 h at 590 nm using an automated microplate reader/stacker (ICN model 340, ICN Biomedicals Inc., Costa Mesa, CA). Absorbance readings were continuously recorded between 1 d and 7 d following inoculation. Microplates were incubated at 22 °C in the dark under a relative humidity approaching 100%. From the string of 96-well absorbance readings for each plate, background-corrected (net) absorbance readings for the 95 substrates at a specified average well color development (AWCD: average of continuous absorbance data) were used for multivariate comparison of sample carbon source utilization patterns (CSUP) (Garland, 1996a). Metabolic diversity or the number of C sources utilized (sum of binary data) was used to compare samples that differed widely. Principal components analysis (PCA) was the multivariate ordination technique used on the correlation matrix of the CSUP with no rotation of factors (Systat 5.2.1, Systat Inc. Evanston, IL). One-way ANOVA testing was performed to determine statistically-significant differences ($p < 0.05$) in factor scores for different locations at a given site or between treatments.

2.2 Tank Farm

The fuel oil tank farm is located at the Idaho National Engineering Laboratory (INEL) in southeast Idaho (USA) in a high desert characterized by sagebrush-steppe vegetation. Three tanks are contained within the 20 x 30 m bermed pad of sandy loam backfill overlying a clay liner; one tank has leaked diesel fuel into the surrounding soils (Fig. 1). Enhanced bioremediation has been chosen to reduce the TPH concentrations that range from < 50 ppm to over 47,000 ppm in the affected soils. Seasonal treatment (May - Oct) consists of weekly tilling, watering and amendments of N and P. Soil samples were aseptically collected at 12 cm and 38 cm depth across a ten-point grid with unaffected fill outside the bermed area used for a control. Ten additional sites 30 meters south of the tank farm in undisturbed sagebrush-steppe terrain were used as native soil controls. Samples were collected and analyzed for CLPP and TPH at May 95, Oct. 95, and Oct. 96. TPH was measured by gas chromatography (modified EPA 8015). A study to investigate the effect on CSUP of storage at 4 °C of soil samples prior to CLPP analysis was conducted on six samples: three from the heavily contaminated area and 3 from the sagebrush terrain outside the tank farm. Splits of the six samples were analyzed the day of collection (Oct. 95) and then five months later and the resulting CSUP examined by PCA.

Fig. 1. Plan view of tank farm and extent of diesel contamination. Sites A-J are within the bermed area (border of figure) and received treatment while control site X (not shown) is located in backfill just outside the bermed area and control sites K-T (not shown) are 30m south in undisturbed soils. Area within the berm is 20x30 meters.

2.3 Snake River Plain Aquifer

The Snake River Plain Aquifer (SRPA) is located in southeast Idaho and consists of thick sequences of fractured basalt flows and thinner interbedded sediments. The aquifer flows northeast to southwest and represents a major source of water for municipal, agricultural and industrial purposes in southern Idaho. The INEL, a U.S. Department of Energy facility, is situated at the northern end of the aquifer, and includes Test Area North (TAN) complex, a collection of experimental and support facilities used for research and development of nuclear reactor performance. Groundwater contamination at TAN has resulted from previous injection well disposal of liquid wastes to the Snake River Plain Aquifer (Fig. 2). Wastes that were disposed via injection well (TSF-05) from 1953 - 1972 included organic, inorganic, and low-level radioactive wastewaters added to industrial and sanitary wastewater. Depth to groundwater in TSF-05 is approximately 63 m bls.

The TSF-05 injection well has been shown to act as the secondary source for a trichloroethylene (TCE) plume that presently extends 3,048 m downgradient with concentrations ranging from 160,000 µg/L at the injection well to less than 5 µg/L at the distal end of the plume. Migration of co-contaminants such as sewage and radionuclides is thought to be limited to several hundred meters from the injection site. Except in the region immediately surrounding TSF-05, the groundwater in the SRPA is saturated with oxygen, has organic carbon values less then 1 mg/l, a pH close to 8.0 and temperatures of 10 °C to 15 °C. While pump and treat remediation is currently underway at this site, in situ technologies under consideration for more efficient remediation of the TAN TCE plume include (a) augmented bioremediation and (b) natural attenuation. Groundwater was collected into sterile containers from wells by use of dedicated, submersible pumps.

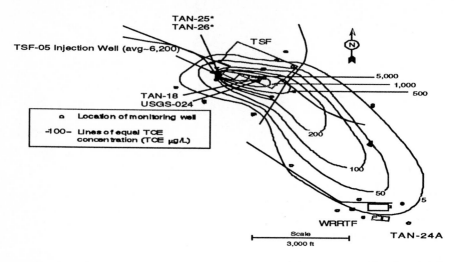

Fig. 2. Plan view of the TAN site showing wells and TCE isopleths (modified from DOE, 1995)

2.4 Iron Creek

Iron Creek (Fig. 3) is located in the Salmon River mountain range, Salmon district of the Salmon National Forest, in central Idaho (latitude: 44°57'30", longitude: 114°07'30"). It is a second order perennial stream with an approximate discharge of 6 ft^3 per second and is located in an area of interest for mineral exploration for metals including copper and cobalt (Erdman and Modreski, 1984). At the location sampled, three abandoned mine adits produce seasonal, circumneutral surface drainage containing metals which flows into the creek while tailings existing below each portal contribute subsurface leachate to the stream. The amount of dilution of Iron Creek by the combined mine drainages is less than 1% (Elton Modroo, Idaho Dept. of Environmental Quality, personal communication).

A transect of six locations in the stream spanning 0.5 km above to 1 km below the point receiving mine drainage were sampled. Triplicate grab samples of water were collected in sterile centrifuge tubes. Dissolved metal concentrations from the Iron Creek and mine drainage waters are from filtered (0.45 μm), acidified samples analyzed by inductively-coupled plasma emission spectroscopy using

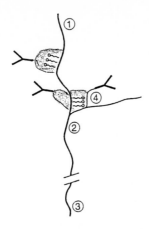

U.S. EPA method 200.7. Periphyton samples were collected in triplicate from 30.7 sq cm areas of in-stream rocks by using a constructed vacuum aided by mechanical brushing. The periphyton suspension was then centrifuged at 1000 x g and washed and resuspended in phosphate-buffered saline (pH 7) to remove bacteria associated with the stream water.

Fig. 3. Relative sampling locations at Iron Creek: 1 = upstream reference sites; 2 = site receiving mine drainage; 3 = downstream recovery sites; 4 = mine drainage

3. Results

3.1 Tank Farm

PCA comparison of the tank farm samples taken before remedial activities in May '95 showed an effect on CSUP due to the presence of TPH compared to control samples of fill dirt outside the bermed area and native soils ($p < 0.001$, 1-way ANOVA of Factor 2 scores, n = 16; data not shown). In Oct. '95 after one season of treatment, TPH concentrations were not detectably reduced. Tilling and watering were not consistently applied during this first season of treatment due to logistical problems. In particular, tilling was not thorough and the 38 cm depth was disregarded for further analysis. Differences were seen in the CSUP from 12 cm deep soils within the spill area from May '95 to Oct. '95, however, these shifts in CSUP were mirrored by shifts in CSUP from control communities (data not shown). The effect of storage at 4 °C on the CSUP of six samples analyzed in Oct. '95 and then re-analyzed five months later were dramatic (Fig. 4). Samples from both contaminated and uncontaminated locations were significantly altered after storage (Factor 1 scores: $p = 0.002$, paired t-test).

Samples that initially exhibited some degree of variation converged on a very similar profile upon storage with the exception of sample location "T". Evaluation of the effectiveness of bioremediation of TPH at the tank farm site during May to Oct. '96 is pending October sampling trip.

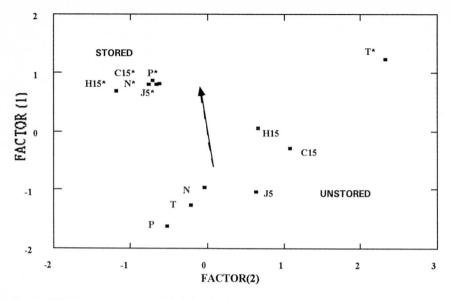

Fig. 4. PCA factor score plot of CSUP of soil samples analyzed on day of collection and then five months later. Arrow shows direction of change in CSUP of the soil communities

3.2 Snake River Plain Aquifer

CLPP were generated from SRPA groundwater from 22 wells inside and outside the contaminant plume at concentrations of TCE up to 15,000 ppb. PCA of the entire dataset of CSUP from the groundwater communities (normalized to AWCD = 0.60) revealed a scattering of samples with the only reliable separation occurring between wells close to the injected site and wells unaffected by the contaminant plume (data not shown). The 22 wells were not arrayed in an ordered transect or grid across the plume as they had been drilled for various purposes over the last 40 years. Further, wells differed in diameter, construction, depth of screened interval and type of pump used to obtain samples. PCA conducted on a set of 3 wells with similar construction, etc. was used to evaluate differences in groundwater communities from "endmember" locations with respect to contaminants including TCE. These wells are the injection well, TSF-05, a well 5 meters downgradient, TAN-25, and a well downgradient past the southern end of the plume, TAN-24A (Fig. 2). PCA grouped independent replicate samples from the two contaminated wells closely, while discriminating the CSUP from the unaffected location, TAN-24A (Fig. 5). Factors 1 and 2 captured 35.0% and 16.2% of the total variance in the dataset, respectively. Main effects of well sampled on Factor 1 scores was determined by 1-way ANOVA (p < 0.001). Tukey's post-hoc test were significant (p < 0.005) for differences in CSUP between the downgradient TAN-24A communities and either TSF-05 or TAN-25. No significant difference was seen between wells TSF-05 and TAN-25 (p = 0.194).

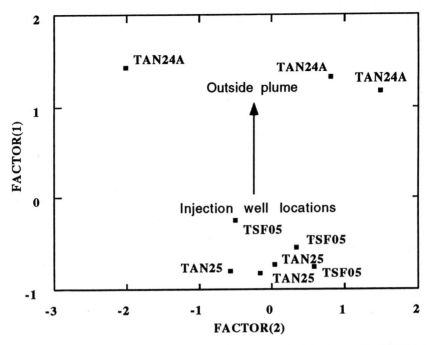

Fig. 5. PCA on TAN groundwater communities. TSF-05 is injection well; TAN-25 is proximal to injection well; TAN-24A is outside of TCE plume. Arrow indicates potential recovery path.

3.3 Iron Creek

Only dissolved iron (0.012 mg/l) and zinc (0.412 mg/l) were above detection (As, Cd, Cr, Co, Cu, Fe, Mn, Mo, Ni, and Zn were measured) at reference sites above the mine drainage while immediately below these inputs, dissolved copper (0.053 mg/l) and manganese (0.008 mg/l) were also found in Iron Creek. At the furthest downstream "recovery" site, manganese values were again below detect in the stream and copper values had dropped by about 50% (0.029 mg/l). Mine drainage surface waters themselves contained measurable dissolved cobalt (0.58 mg/l), copper (0.064 mg/l), manganese (0.381 mg/l), zinc (0.851 mg/l) and iron (0.211 mg/l). Due to the high variance and values of replicate sample zinc values (data not shown), it is suspected that zinc concentrations may reflect a laboratory artifact.

At the two reference sites upstream of the mine drainage input, bacterioplankton communities had positive response to 81 - 89 of the carbon sources in the Biolog GN microplates. Periphyton communities respired 77 - 91 substrates per microplate at the upstream reference locations. Averaging the triplicate samples from both of the reference sites and using this response as

100%, the percent of the carbon sources (normalized to reference locations) utilized by communities in Iron Creek downstream of the mine drainage and from mine drainage are shown in Figure 6. The Iron Creek site immediately downstream of the mine drainage input respired only 35% (periphyton) and 58%

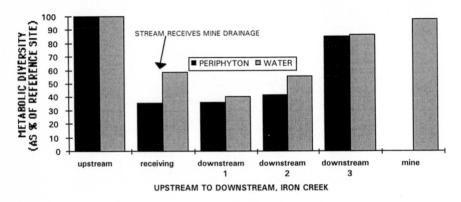

Fig. 6. Substrate utilization by Iron Creek communities along stream gradient. Datum are expressed as mean of three independent replicate samples

(water) of the substrates used by upstream reference communities. At three subsequent downstream sites that cover the next 1 km of the stream, some recovery is seen in the number of substrates utilized until at the farthest downstream site, 85% of the reference response is exhibited by both water and periphyton communities (Fig. 6). Mine surface drainage communities oxidized 81 - 89 carbon sources in independent triplicate samples.

Principal components analysis (PCA) was performed on the CSUP of communities that achieved an AWCD of 0.5 which was used to normalize for differences in extent/rate of color development. These samples consisted of water and periphyton communities from the two reference upstream sites, the farthest downstream site and the mine drainage surface waters. One of three replicates for the upstream site 1 water, upstream site 2 periphyton and mine drainage site were anomalous due to laboratory procedural errors and were removed from the dataset prior to PCA to prevent domination of this variance-based ordination method of sample discrimination. A plot of Factor 1 versus Factor 2 scores for Iron Creek communities is shown in Figure 7. PCA Factors 1 and 2 captured 31.1% and 19.3% of the total variance in the dataset, respectively. Both periphyton and water from the upstream reference sites exhibited similar CSUP while mine drainage and downstream sites showed varying degrees of dissimilarity from the reference communities. Since PCA factor scores for water and periphyton upstream reference sites were very similar, they were treated as a group and tested for significantly different CSUP against downstream waters, periphyton and mine drainage sites by 1-way ANOVA. Main effects of site location on both Factor 1 and Factor 2 scores were significant ($p < 0.001$). Tukey's post-hoc test demonstrated significant differences in Factor 1 scores for: reference upstream sites vs. mine drainage ($p < 0.001$), downstream periphyton ($p < 0.001$), and downstream water ($p < 0.05$); downstream water vs downstream periphyton ($p = 0.001$) and mine drainage ($p = 0.001$). Tukey's post-hoc test on Factor 2 scores was significant ($p < 0.001$) for all pairs of sites except reference upstream sites vs. downstream waters.

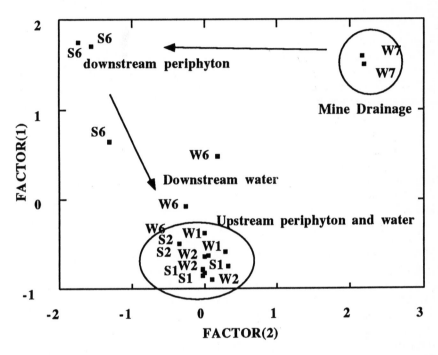

Fig. 7. PCA of Iron Creek communities. M7 = mine drainage; S6 = downstream periphyton; W6 = downstream water; W1 & W2 = upstream reference waters; S1 & S2 = upstream reference periphyton. Identical numbers are independent replicate samples from same location.

4. Discussion

While multivariate analysis of CSUP easily distinguished between tank farm soil communities at sites with high TPH values and unaffected control locations during the initial (May '95) sampling, that result was not unexpected (data not shown). Sampling after one season of remedial treatment failed to show any reduction in TPH values, probably due to logistical problems with the tilling and watering aspects of the treatment regime as well as an exceptionally cool summer in southeast Idaho. CSUP from the contaminated areas did shift from May to Oct. '95 but did not become more similar to control areas sampled in Oct. '95 (data not shown). Moreover, the shift in CSUP of the contaminated areas was paralleled by a shift in the CSUP of control site communities. Based on the above observations and ineffectiveness of the '95 season treatment in reducing TPH, CSUP shifts from May to Oct. '95 were attributed to seasonal affects.

The effect of sample holding time was investigated and found to have a significant effect on the CSUP of soil communities, regardless of degree of TPH values (Fig. 4). Interestingly, both uncontaminated and contaminated soils converged on a similar CSUP after five months storage at 4 °C indicating strong selection during storage. This finding emphasizes the importance of rapid CLPP analysis following collection of samples.

Due to the lack of accepted standards for ecologically-relevant environmental monitoring in soils, the potential of microbial community profiling methods is greatest in terrestrial systems. However, the factors contributing to natural spatial variability in microbial communities need to be accounted for in realizing effective contaminant monitoring. Typical factors that need to be considered in sampling designs at contaminated sites include soil texture and moisture, depth of sampling, location and type of vegetation, and slope aspect. Temporal changes including temperature, irradiation, and their effects on abiotic and biotic variables above need to be incorporated into the sampling regime. Where successful control of many of the primary variables listed can be achieved, maximizing the number of independent replicate samples should allow distinction of contaminant effects on microbial communities from other sources of variation.

At the SRPA site, CLPP was only able to reliably distinguish between groundwaters which represent endmembers with respect to the plume of TCE (Fig. 5). This site in particular illustrates many of the difficulties and considerations for application of CLPP to environmental monitoring. In the highly fractured porous media of the SRPA, a number of environmental variables need to be accounted for to detect the impact of contaminants on CLPP. Some of the natural variables that affect the CLPP of the SRPA communities sampled are the groundwater geochemistry as it reflects local lithology, mineralogy and site and distance of recharge area, hydrological parameters (e.g., transmissivity) and physical variables such as the porosity and cation-exchange capacity of aquifer matrix. Other variables are associated with the accessing samples such as the general location of wells, their location, depth to screened interval and type of pump used to acquire water. Even pumping rates and duration may affect to composition of the microbial communities from a given well (Thomas et al., 1987). Complicating the dissection of toxic effects on microbial communities at the TAN site is the presence of exogenous growth substrates (e.g., sewage) that was also pumped down the injection well. It is thought that the components of the sewage are limited in their migration with respect to TCE (DOE, 1995), probably through degradation and sorption, so selective enrichment of microbial communities from this source may be of little concern downgradient within the TCE plume.

Discriminating effects of contaminants on the CLPP of groundwater communities from the effects of other variables is particularly difficult at this heterogenous site where the access to aseptic samples is limited and the existing set of wells are not optimally placed or constructed. Vertical and horizontal distributions of contaminants need to be considered. Colwell and Lehman (1997) found that there was little difference in the CLPP of SRPA groundwater communities in a vertical transect within one borehole (relatively uncontaminated) where hydraulic conductivities varied by orders of magnitude, although the result at a contaminated location may be different. Besides concerns with the number of wells and their locations, the SRPA highlights the requirement for collateral data on the physical and chemical characteristics of the groundwater samples for interpreting CLPP and the need to collaborate with workers in other disciplines, e.g., geochemistry and hydrology. One path to boosting the 'signal-to-noise ratio' of the CLPP data is to convert the continuous data to binary form prior to analysis, although definition of a threshold for a positive response across differing samples is difficult. Non-parametric methods for calculation and comparison of 95% confidence intervals of treatment groups may be the most appropriate analysis choice for some datasets (Bloom, 1980).

At the Iron Creek site, CLPP was very effective at detecting the impact from a small input of mine drainage on stream communities. At the site immediately downstream of the mine drainage and two subsequent downstream sites, simply examining the number of substrates used by either the bacterioplankton or periphyton was sufficient to determine impact (Fig. 6). The periphyton communities were more severely inhibited compared to the seston communities, a trend that remained until the site farthest from the mine drainage were the two communities utilized a similar number of substrates. Dissolved metals that were input to the stream from the mine drainage probably were removed from the water column of the stream flowpath by precipitation, sorption and uptake by benthic algae. The relative response of the planktonic and sessile communities is consistent with metals being removed from the water column and becoming concentrated in the stream periphyton, a phenomena that has been observed at this site (Erdman and Modreski, 1984). In comparing communities from the mine drainage and farthest downstream site to the upstream reference communities, it was necessary to use multivariate statistics to distinguish between CSUP of these communities since they all utilized roughly the same number of substrates (Figs. 6 & 7). ANOVA testing of Factor 1 and 2 PCA scores easily distinguished the mine communities ($p < 0.001$) and downstream periphyton communities from the upstream reference communities (Fig. 7). Factor 1 PCA scores for the downstream communities were also significantly different ($p < 0.05$) from the upstream reference communities showing that despite similar number of substrates used, the CSUP (community structure?) of these communities was still different. The plot of factor scores suggest that the downstream water bacterioplankton communities were in the process of recovering to the point of exhibiting CSUP similar to the upstream reference sites while the periphyton communities were either at earlier point in their recovery or were more permanently restructured into a different but functionally capable community like the mine drainage. In a parallel study on the benthic algae at these stream locations, a shift in relative species abundance was seen below the mine drainage and continued to the farthest downstream site sampled with a progressive trend toward community structures resembling upstream reference locations (R. Genter, personal communication).

Advantages of using the CLPP approach for environmental assessment are that CLPP integrates multiple populations of heterotrophs that are indigenous to the site of interest. Measurement of mixed communities dampens the variability due to distribution of individual populations and mitigates the significance of choosing one population with its inherent tolerances. Multivariate measures (i.e., CLPP) of community profile allow greater resolution of differences in samples than univariate measures (i.e., biomass or glucose turnover). Community profiles also may allow more subtle community restructuring due to chronic presence of contaminant to be distinguished from more gross impacts on the net physiological process of the microbiota (Lehman and Garland, unpublished data). Bioassays based on indigenous communities allow more accurate characterization of contaminant impact that reflects in situ bioavailability of chemicals to native organisms. Alternative methods of multivariate community profiling are represented by a suite of molecular approaches (e.g., Sayler et al., 1992) and whole community ester-linked phospholipid fatty acid profiles (e.g., Tunlid and White, 1992). Molecular and fatty-acid based methods of assessing microbial community structure are not growth-dependent and therefore may include a greater proportion of native populations that are understood to be non-culturable. On the other hand, technical difficulties with extraction of nucleic acids (often at low levels) from environmental samples and the expense associated with molecular community profiling limit the immediate application of these methods for environmental assessment. Phospholipid fatty acid (PLFA) community profiles have been demonstrated as useful in evaluation environmental perturbation (Guckert et al., 1992), however, the expertise and expense required for these analyses exceed that of CLPP. In unconcentrated groundwater and surface water samples, Lehman et al. (unpublished data) have detected an sufficient CLPP signal while sample splits for PLFA analysis were below the method detection limit.

The primary limitation of the CLPP approach is its dependence on cell growth (or activity) and the resulting selectivity, e.g., sampling, extraction, nutrient and incubation conditions. The CLPP approach empirically defines some segment of the natural community and consistent performance of a standard technique on a number of independent replicate samples may allow useful applications in environmental assessment. The focus should be on discriminating contaminant effects from other sources of variation on the CLPP signal and quantitatively demonstrating recovery of the CLPP of contaminated sites to that of an appropriate reference site. Due to the volume of data generated with near-continuous monitoring of color development in the microplates, automation and associated validation of data acquisition, reduction and analysis protocols can provide effective characterization in a minimum of time.

5. Acknowledgements

Funding for research and travel received from the U.S. Department of Energy Office of Energy Research, Subsurface Science Program, Dr. Frank Wobber, program manager. Research was supported in part by Lab-Directed Research and Development funding at the Idaho National Engineering Laboratory operated under contract DE-AC-07-94ID-13223 by Lockheed-Martin Idaho Technologies Co. Collaboration and technical assistance is acknowledged from Dr. Bob Genter, Brady Lee, Brad Blackwelder, Don Maiers, Callie Talbert, Joe Lord and Steve Ugaki.

6. References

Bloom SA (1980) Multivariate quantification of community recovery. In: Cairns J (ed) The Recovery Process in Damaged Ecosystems, Ann Arbor Science Publications, Ann Arbor, pp. 141-151

Colwell FS, Lehman RM (1997) Microbial communities from hydrologically-distinct zones of a basalt aquifer. Microb Ecol (in press)

Ellis RJ, Thompson IP, Bailey MJ (1995) Metabolic profiling as a means of characterizing plant-associated microbial communities. FEMS Microbiol Ecol 16: 9-18

Erdman JA, Modreski PJ (1984) Copper and cobalt in aquatic mosses and stream sediments from the Idaho cobalt belt. J Geochem Exploration 20: 75-84

Garland JL, Mills AL (1991) Classification and characterization of heterotrophic microbial communities on the basis of community-level sole carbon source utilization. Appl Environ Microbiol 57: 2351-2359

Garland JL, Mills AL (1994) A community-level physiological approach for studying microbial communities. In: Ritz K, Dighton J, Giller KE (eds) Beyond the Biomass: Compositional and Functional Analysis of Soil Microbial Communities, John Wiley, Chichester, pp. 77-83

Garland JL (1996a) Analytical approaches to the characterization of samples of microbial communities using patterns of potential carbon source utilization. Soil Biol Biochem 28: 213-221

Garland JL (1996b) Patterns of potential carbon source utilization by rhizosphere communities. Soil Biol Biochem 28: 223-230

Guckert JB, Nold SC, Boston HL, White DC (1992) Periphyton response in an industrial receiving stream: lipid-based physiological stress analysis and pattern recognition of microbial community structure. Can J Fish Aquat Sci 49: 2578-2587

Haack SK, Garchow H, Klug MJ, Forney LJ (1995) Analysis of factors affecting the accuracy, reproducibility and interpretation of microbial community carbon source utilization patterns. Appl Environ Microbiol 61: 1458-1468

Insam H, Amor K, Renner M, Crepaz C (1996) Changes in the functional abilities of the microbial community during composting of manure. Microb Ecol 31: 77-87

Joern A, Hoagland KD (1996) In defense of whole-community bioassays for risk assessment. Environ Toxicol Chem 15: 407-409

Lehman RM, Colwell FS, Ringelberg DB, White DC (1995) Combined microbial community-level analyses for quality assurance of terrestrial subsurface cores. J Microbiol Meth 22: 263-281

Sayler GS, Nikbakht K, Fleming JT, Packard J (1992) Applications of molecular techniques to soil biochemistry. In: Stotzky G, Bollag J-M (eds) Soil Biochemistry (Vol 7), Marcel Dekker, New York, pp. 131-172

Thomas JM, Lee MD, Ward CH (1987) Use of groundwater in assessment of biodegradation potential in the subsurface. Environ Toxicol Chem 6: 607-614

Tunlid A, White DC (1992) Biochemical analysis of biomass, community structure, nutritional status, and metabolic activity of microbial communities in soil. In: Stotzky G, Bollag J-M (eds) Soil Biochemistry (Vol 7), Marcel Dekker, New York, pp. 229-262

U.S. Environmental Protection Agency (1989) Ecological Assessment of Hazardous Waste Sites: A Field and Laboratory Reference. EPA/600/3-89/013, U.S. Environmental Protection Agency, Washington, DC

U.S. Environmental Protection Agency (1994) Interim guidance on determination and use of water-effect ratios for metals. EPA-823-B-94-001, U.S. Environmental Protection Agency, Washington, DC

U.S. Department of Energy (1995) Record of Decision. Declaration of Technical Support Facility injection well (TSF-05) and surrounding groundwater contamination (TSF-23) and miscellaneous no action sites final remedial action. Operable Unit 1-07B. Waste Area Group 1, Idaho National Engineering Laboratory, Idaho Falls

Winding AK (1994) Fingerprinting bacterial soil communities using Biolog microtiter plates. In: Ritz K, Dighton J, Giller KE (eds) Beyond the Biomass: Compositional and Functional Analysis of Soil Microbial Communities, John Wiley, Chichester, pp. 85-94

Carbon Source Utilization by Microbial Communities in Soils under Organic and Conventional Farming Practice

Andreas Fließbach and Paul Mäder

Research Institute of Organic Agriculture (FiBL), Ackerstrasse, CH-5070 Frick,
E-mail: fliessbach@fibl.ch

Abstract. In a long-term field trial in which organic and conventional agricultural systems were compared since 1978 we analyzed soil microbial biomass, microbial activity and substrate utilization patterns by the Biolog GN microplates. Microbial biomass and the C_{mic}-to-C_{org} ratio was distinctly higher in organic plots whilst the metabolic quotient qCO_2 as an indicator of the energy requirement of soil microorganisms was lower. Substrate utilization profiles were affected by the different long-term treatments, but also indicated differences of short-term effects like crop and soil management steps.

Keywords. Organic farming, microbial biomass, C_{mic}-to-C_{org} ratio, qCO_2, substrate utilization patterns, functional diversity

1. Introduction

Growing concern about the environmental and socio-economic effects of conventional agricultural practice has led farmers, consumers and recently politicians to seek alternative approaches, that are more sustainable in the view of environmental safety, soil protection, and the socio-economic security of the farmer. Especially in Western European countries, Australia, New Zealand and North America farmers developed organic farming systems to maintain soil fertility and profitability without the use of synthetic pesticides and fertilizers (Reganold 1995). Organic farming intends to provide practices that may keep the agricultural system close to natural systems.

Natural ecosystems are typically in a steady state as determined by environmental and geological factors. However agricultural systems, with crop and plant residue removal and intensive soil tillage will hardly reach this state. According to Odum's (1969) hypothesis climax communities support the highest possible biodiversity, with numerous species interactions, and a balanced biomass production to respiration ratio, as well as a low nutrient loss. The hypothesis was recently confirmed by Tilman *et al.* (1996), who found a reduced nitrate leaching potential of soils with a high plant diversity.

For soil ecosystems, Odum's theory of bioenergetics in ecosystem development was confirmed by studies of Insam and Domsch (1988) who found a decreasing CO_2 release per unit microbial biomass (qCO_2) from a series of soils from young to mature sites. Anderson and Domsch (1989; 1990) found that the qCO_2 of the soil microbial biomass (C_{mic}) in soils under crop rotation was significantly lower than in monoculture soils and the ratio of C_{mic} to total soil organic C was significantly higher. It may thus be assumed that a diverse plant community will favour effective soil microbial communities by reducing their energy demand.

Under similar environmental conditions the only explanation for these qualitative differences between communities is a different species composition that may use the available C sources more efficiently. However, only a few researchers have attempted to link these two ecological domains of autecology and synecology (Insam et al. 1996).

The present paper addresses functional properties of microbial communities from soils of different agricultural systems. A recent method based on the utilization patterns of 95 different C sources on a microtiter plate was tested. It has already been applied to soils and was able to distinguish between different sites and treatments (Bossio and Scow 1995; Garland and Mills 1991; Winding 1993; Zak et al. 1994). It was the aim of our investigations to see if a) microbial populations from soils of different agricultural systems differ in their community energetics and b) if differences in community energetics can be correlated to the functional abilities of the soil microbial community.

2. Materials and Methods

2.1 Soils and field experiment

Investigations were carried out on soils from plots of a long-term field trial (DOC) at Therwil (CH), that was started in 1978 to compare different agricultural systems regarding system productivity, nutrient balance, energy efficiency, soil fertility, and in general their effect on the environment (Besson and Niggli 1991). The soil is a luvisol on deep deposits of alluvial loess that has been cultivated as an arable soil for a long time. Long-term annual mean temperature is 9.0 °C and annual rainfall averages at 872 1 m^{-2} (310 m above sea-level). A seven year crop rotation included potatoes, winter wheat 1, beet roots, winter wheat 2, and three years of grass-clover in all systems. Our investigation included the following agricultural systems, where the amount of organic fertilizer corresponded to 1.4 livestock units ha^{-1} which resulted in a mean annual amount of 2 mg organic matter ha^{-1} in all organically fertilized treatments:

unfertilized: Unfertilized since 1978, but amended with bio-dynamic preparations.

bio-dynamic: Fertilized with aerobically composted farm-yard-manure (FYM) (C/N=8) and amended with mineral and herbaceous preparations according to biodynamic farming practice (Kirchmann 1994).

organic: Fertilized with slightly aerobically rotted FYM (C/N=11) according to the organic farming practice in Switzerland.

conventional: Fertilized with stacked, anaerobically rotted FYM (C/N=12) and an additional mineral fertilization according to the official Swiss extension service, integrated pest management.

mineral NPK: Unfertilized from 1978 until 1985, but then fertilized with NPK according to the official extension service, integrated pest management.

We investigated soils from winter wheat plots. Wheat has been cultivated in two consecutive years in two parallel crop rotation subunits of the trial. Soils from 0-20 cm were sampled in March 1995 and March 1996 at tillering and two months after the harvest of winter wheat in October 1995 as bulked samples from each of the four field replicates. In 1995 wheat followed red beets and in 1996 potatoes in the crop rotation. In the laboratory, soils were kept at 4 °C until they were sieved (2 mm) and water adjusted to 50 % water holding capacity (ca. 24 % H_2O of dry matter).

Data on soil pH (0.1 M KCl, 1/10; weight/vol) organic C and N (CHN analyzer, LECO) are given in table 1.

Table 1. pH, organic carbon (C_{org}), total N (N_t) and C/N ratio in soils from the DOC field trial (n=4). Different letters indicate significant differences at p=0.05

	pH (0.1 M KCl)	C_{org}	N_t	C/N
unfertilized	5.1 ab	1.32 a	0.140 a	9.58 a
bio-dynamic	5.7 a	1.61 b	0.169 b	9.56 a
organic	5.4 a	1.34 ab	0.148 ab	9.11 a
conventional	5.2 ab	1.37 ab	0.132 ab	10.36 a
mineral NPK	4.9 b	1.41 ab	0.156 ab	9.18 a

2.2 Soil microbial biomass

Soil microbial biomass carbon (C_{mic}) was estimated by chloroform-fumigation-extraction (CFE) according to Vance *et al.* (1987). CFE was done on 25 g subsamples that were extracted with 100 ml of a 0.5 M K_2SO_4 solution. Total organic carbon (TOC) in soil extracts was determined by infrared spectrometry after combustion at 850 °C (DimaTOC, Dimatec, Essen). Soil microbial biomass was then calculated according to the formula:

$C_{mic} = E_C/k_{EC}$

EC= (TOC in fumigated samples - TOC in control samples)

$k_{EC} = 0.45$ (Joergensen 1995; Martens 1995)

2.3 Carbon source utilisation assay

Soils were preincubated at 22 °C for 5 days. After comparing different extractants and dilutions we decided to use the extraction procedure that gives the highest values for average well colour development (AWCD) and the lowest variability among repeated measurements. 10 g (dry matter) of soil was suspended in 100 ml of extractant on a rotary shaker at 300 rev min^{-1} for 30 min. Soil suspensions, prepared with a 0.9 % NaCl-solution showed higher values and lower variation than with 0.1 % sodium hexametaphosphate, where an addition of $CaCl_2$ was necessary to precipitate dispersed clay minerals (Haack *et al.* 1995). NaCl extracts were allowed to settle for 10 min to clear the supernatant, which was diluted tenfold to obtain a final dilution of 10^{-2}.

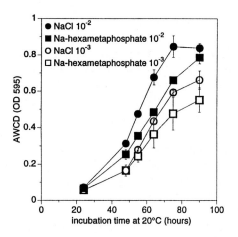

Fig. 1. Average well colour development in Biolog microplates as affected by the solution used for extraction of microorganisms and by the dilution of the soil suspension (n=3).

This suspension (125 µl per well) was directly inoculated to Biolog GN microplates (Biolog Inc. Hayward, Cal., USA) (Garland and Mills 1991). Microplates were kept at 22 °C until colouration became visible, then the plate absorbance was measured repeatedly (BioRad, Model 450). For data processing we used set points, where the average well colour development reached values between 0.6 and 0.7 absorption units. This was usually the case after 60 to 65 hours of incubation at 22 °C. Data were processed in the following ways:

Raw difference (RD): $X - X_0$ where X is the raw value of each well and X_0 the OD_{595} of the water blank, negative scores were set to zero.

Average well colour development (AWCD): $\Sigma RD/95$

Number of utilized substrates: Number of substrates with RD > AWCD

Diversity was calculated according to population ecology by using raw difference data: - $\sum p_i\ (lnp_i)$ where p_i is the ratio of activity on a particular substrate to the sum of activities on all substrates.

2.4 Statistics

Biolog results were analyzed with CANOCO software from Microcomputer Power, Inc. (Ithaca, N.Y.) (Jongman *et al.* 1995). For diversity calculations substrates were treated as individual species. AWCD transformed data were used to perform principal component analysis (PCA).

3. Results and Discussion

3.1 Soil microbial biomass and activity

The different agricultural systems, represented in the DOC field trial, mainly differing in fertilization and plant protection strategy, resulted in marked differences in soil microbial biomass (C_{mic}) and microbial activity (Mäder *et al.* 1995). After 18 years, C_{mic} in soils of the two organic plots was significantly higher than in conventional soils (Fig. 2). The organic manure treatment in the conventional plots did not cause a significant difference to the mineral fertilizer treatment and the unfertilized plot. Corresponding results were found in system comparison trials at Darmstadt (D) (Bachinger 1995; Raupp 1995) and Järna (S) (Pettersson *et al.* 1992), indicating that organic rather than conventional farming exerts positive effects on the size and activity of the soil microbial biomass (Mäder *et al.* in press).

Although microbial biomass and soil basal respiration were highest in the bio-dynamic treatment, we found the lowest qCO_2 in these plots (Fig. 3), indicating that the microbial populations of these plots utilize the organic substrates more efficiently. The key to understanding differences of microbial metabolic activity among the treatment plots is supposed to lie within the community structure of soil microorganisms.

The ratio of microbial biomass C to total soil organic C was also higher in the organic plots than in the conventional and especially the plots with mineral fertilizer exclusively (Fig. 4). Since this ratio is indicating organic matter quality and availability it suggests that long-term amendments with fertilizers of different quality will result in a change of organic matter pools and that fertilizers are favouring soil organisms to a different extent (Anderson and Domsch 1989).

3.2 Substrate utilization by soil microbial communities

In a first step we attempted to find site differences by the use of univariate measures of substrate utilization patterns. Microbial functional diversity as indicated by the Shannon index showed the same ranking over the whole incubation period of the microplates, but at different levels. Differences due to the DOC-treatments were obtained in samples from March 1995 and March 1996, where the bio-dynamic plots showed highest values and the conventional treatment the lowest (Fig. 5). In samples from October 1995 under emerging grass-clover and two months after the winter wheat harvest differences between the systems became very small. Possibly a different microflora has developed after incorporation of straw and residues, resulting in a more diverse substrate utilization of the microbial communities of all treatments.

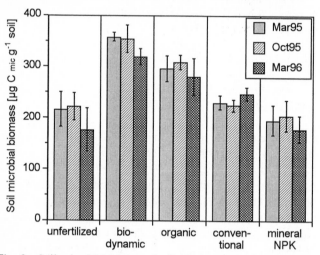

Fig. 2. Soil microbial biomass in field plots of the DOC-experiment at three sampling dates under winter wheat. The columns represent the mean of four field replicates (Standard error bars shown).

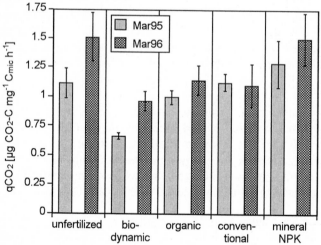

Fig. 3. Metabolic quotient for CO_2 (qCO_2) in plots of the DOC field trial in March 1995 and 1996. Columns are the mean of four field replicates (Standard error bars shown).

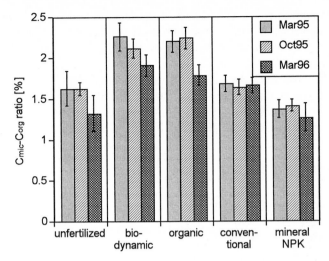

Fig. 4. C_{mic}-to-C_{org} ratio in plots of the DOC field trial in March 1995 and 1996. Columns are the mean of four field replicates (Standard error bars shown).

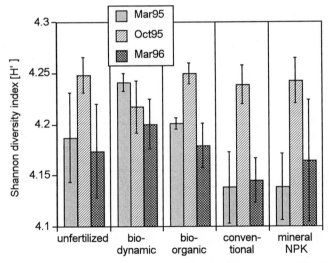

Fig. 5. Microbial functional diversity (Shannon Index) at three sampling dates under winter wheat. The columns are the mean of the four field replicates (Standard error bars shown).

Garland and Mills (1991) and Zak *et al.* (1994) identified different groups of chemically related substrates, of which the carbohydrates, carboxylic acids and amino acids were most numerous. By adapting the approach of Vahjen *et al.* (1995) we counted the number of substrates that showed higher absorption than the average plate absorbance. Differences among the number of these intensely utilized substrates were small, but in most cases due to differences of carbohydrate utilization (Fig. 6). Especially in the March samples substrate utilization was greatly affected by carbohydrates. Carbohydrate utilization in the bio-dynamic soils was 50 % higher than in the conventional, whereas the other substrates did not differ that much. In October the relative proportion of each of the substrate groups was almost the same, showing differences to the conventional treatment of less than 10 %, however mean utilization of carbohydrates over the treatments was markedly higher at this sampling time than in the samples of March.

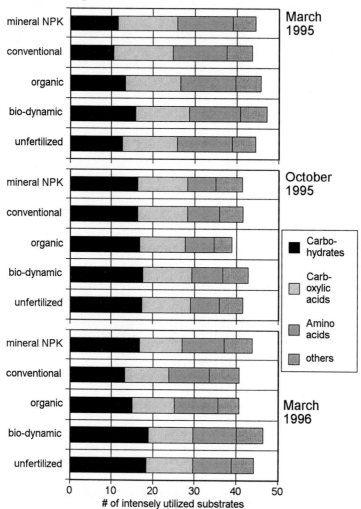

Fig. 6. Number of intensely utilized substrates (absorption > average plate absorbance) among chemically similar substrate classes. The columns represent the mean of four field and three microplate replicates.

Principal component analysis (PCA) applied to the absorption values transformed by the AWCD (Garland and Mills 1991) separated some of the DOC-treatments, by ordination of the first two principal components. The first two principal components accounted for more than 50 % of the total variance. However variation along with field replicates of the same treatments were often higher than variation between them (Fig. 7).

The substrate utilization patterns of microbial communities from the field trial soils showed distinct overlapping. However, in March 1995 the two organic systems showed a close ordination of the first two principal components as did the two conventional systems. In October 1995 the ordination points of the field replicates did not allow for a distinct grouping, but in March 1996 the three organically fertilized plots ordinated close by, whereas the unfertilized and minerally fertilized plots were far apart.

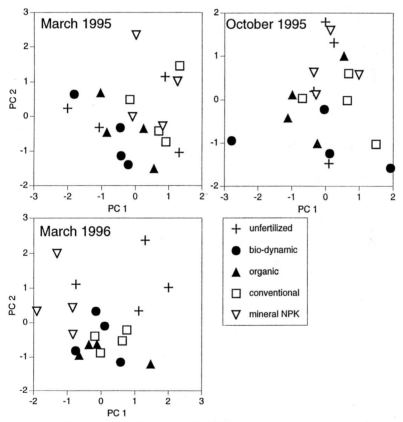

Fig. 7. Ordination of the first two principal components. Sample scores for each plot of the field trial represent the mean of three replicate microplates.

Variation within the field replicates can be explained by inhomogeneities, that were also observed with measurements of microbial biomass and activity.

Especially the subplots that were used in 1996 showed distinct variation among field replicates in the mineral and the unfertilized treatments. Ordination techniques that allow for the inclusion of covariables to account for block effects would be helpful.

Differences between the sampling dates in the same subplot (March and October 1995) are probably due to the soil disturbance after winter wheat cropping. Plant residues have been incorporated and farmyard manure and mineral fertilizers have been applied to the field plots two months prior to sampling. During this period a microflora might have developed that was not specifically adapted to the site, but to the nutrient and organic matter input. Therefore the substrate utilization patterns of the October sample did not differ among treatments.

Soils from different agricultural systems have been shown to differ in soil microbial biomass and activity. Holistic methods for their determination were little affected by sampling date. Substrate utilization reacted more sensitively to short-term management effects like tillage or fertilization rather than the long-term system effects.

The metabolic quotient for CO_2 in the two organic systems was markedly lower than in the two conventional treatments, indicating that the soil microbial population needs less energy for maintenance. A significantly higher diversity in organic soils was only found for the spring sampling in 1995. Hence, the hypothesis that differences due to community energetics are explainable by the functional richness and diversity of the soil microbial community needs further examination.

4. References

Anderson TH, Domsch KH (1989) Ratios of microbial biomass carbon to total organic carbon in arable soils. Soil Biol Biochem 21: 471-479

Anderson TH, Domsch KH (1990) Application of eco-physiological quotients (qCO_2 and qD) on microbial biomasses from soils of different cropping histories. Soil Biol Biochem 22: 251-255

Bachinger J (1995) Effects of organic and mineral fertilizer on chemical and microbial parameters of C- and N-dynamics and root parameters. Mäder P, Raupp J (eds) Effects of low and high external input agriculture on soil microbial biomass and activities in view of sustainable agriculture, 2nd meeting of the EU-concerted action (AIR3-CT94-1940), Oberwil, p 52-58

Besson J-M, Niggli U (1991) DOK-Versuch: Vergleichende Langzeituntersuchungen in den drei Anbausystemen biologisch-Dynamisch, Organisch-biologisch und Konventionell. I. Konzeption des DOK-Versuchs: 1. und 2. Fruchtfolgeperiode. Schweiz Landw Fo 31: 79-109

Bossio DA, Scow KM (1995) Impact of carbon and flooding on the metabolic diversity of microbial communities in soils. Appl Environ Microbiol 61: 4043-4050

Garland JL, Mills AL (1991) Classification and characterization of heterotrophic microbial communities on the basis of patterns of community-level-sole-carbon-source utilization. Appl Environ Microbiol 57: 2351-2359

Haack SK, Garchow H, Klug MJ, Forney LJ (1995) Analysis of factors affecting the accuracy, reproducibility, and interpretation of microbial community carbon source utilization patterns. Appl Environ Microbiol 61: 1458-1468

Insam H, Amor K, Renner M, Crepaz C (1996) Changes in functional abilities of the microbial community during composting of manure. Microbial Ecol 31: 77-87

Insam H, Domsch KH (1988) Relationship between soil organic carbon and microbial biomass on chronosequences of reclamation sites. Microb Ecol 15: 177-188

Joergensen RG (1995) Die quantitative Bestimmung der mikrobiellen Biomasse in Böden mit der Chloroform-Fumigations-Extraktions-Methode. Göttinger Bodenkundliche Berichte, Universität Göttingen, Göttingen,

Jongman RHG, Ter Braak CJF, Van Tongeren OFR (1995) Data Analysis in Community and Landscape ecology. 2/Ed. Cambridge University Press, Cambridge

Kirchmann H (1994) Biological dynamic farming - an occult form of alternative agriculture? J Agric Environ Ethics 7: 173-187

Mäder P, Fließbach A, Wiemken A, Niggli U (1995) Assessment of soil microbial status under long-term low input (biological) and high input (conventional) agriculture. Mäder P, Raupp J (eds), Effects of low and high external input agriculture on soil microbial biomass and activities in view of sustainable agriculture, 2nd meeting of the EU-concerted action (AIR3-CT94-1940), Oberwil (CH), p 24-38

Mäder P, Pfiffner L, Fließbach A, von Lützow M, Munch JC (in press) Soil ecology - The impact of organic and conventional agriculture on soil biota and its significance for soil fertility. Oestergaard TV (ed), Fundamentals of Organic Agriculture, Proc. of the 11th IFOAM International Scientific Conference, Copenhagen, DK

Martens R (1995) Current methods for measuring microbial biomass C in soil: Potentials and limitations. Biol Fertil Soils 19: 87-99

Odum EP (1969) The strategy of ecosystem development. Science 164: 262-270

Pettersson BD, Reents HJ, von Wistinghausen E (1992) Düngung und Bodeneigenschaften. Ergebnisse eines 32-jährigen Feldversuches in Järna, Schweden. Institut für biologisch-dynamische Forschung, Darmstadt

Raupp J (1995) The long-term trial in Darmstadt: Mineral fertilizer, composted manure and composted manure plus all bio-dynamic preparations. Raupp J (ed), Main effects of various organic and mineral fertilization on soil organic matter turnover and plant growth, 1st meeting of the EU-concerted action (AIR3-CT94-1940), Darmstadt, p 28-36

Reganold JP (1995) Soil quality and profitability of biodynamic and conventional farming systems: A review. Amer J Alternative Agric 10: 36-45

Tilman D, Wedin D, Knops J (1996) Productivity and sustainability influenced by biodiversity in grassland ecosystems. Nature 379: 718-720

Vahjen W, Munch JC, Tebbe CC (1995) Carbon source utilization of soil extracted microorganisms as a tool to detect the effects of soil supplemented with genetically engineered and non-engineered *Corynebacterium glutamicum* and a recombinant peptide at the community level. FEMS Microbiol Ecol 18: 317-328

Vance ED, Brookes PC, Jenkinson DS (1987) An extraction method for measuring soil microbial biomass C. Soil Biol Biochem 19: 703-707

Winding A (1993) Fingerprinting bacterial soil communities using Biolog microtiter plates. In: Ritz K, Dighton J, Giller KE (eds) Beyond the Biomass: compositional and functional analysis of soil microbial communities, Vol. , John Wiley and Sons Ltd., Chichester, UK, pp 85-94

Zak JC, Willig MR, Moorhead DL, Wildman HG (1994) Functional diversity of microbial communities: a quantitative approach. Soil Biol Biochem 26: 1101-1108

Characterization of Microbial Communities in Agroecosystems

Ann C. Kennedy[1] and Virginia L. Gewin[2]

[1]USDA-ARS Pullman, WA 99164-6421,
[2]Washington State University Pullman, WA 99164-6420

Abstract. Substrate utilization and stress response patterns were assessed for bacteria, fungi, and actinomycetes in both wheat and prairie systems. Landscape position was also a potential source of variation among the samples. The data were further analyzed according to distinct isolate groupings. Bacterial isolates from the tilled systems had higher carbon source utilization as compared to prairie isolates, while actinomycetes isolates from the prairie system had higher carbon sources utilization as compared to the tilled system isolates. Overall, actinomycetes isolate responses were higher for both the tilled and prairie systems. Actinomycetes isolates also had the highest growth rates in the presence of stressors for both systems as compared to bacterial and fungal isolates. Microbial substrate utilization differed with landscape position. The isolates from the tilled system had higher substrate utilization from foot slope, while the isolates from the prairie system had higher substrate utilization responses from side slope. Ridge top responses were similar. Relationship analyses among the substrates, isolates, and management systems support these findings. Actinomycetes had strong associations with arginine, sorbitol, xylan, streptomycin, and penicillin. Investigations of this nature can provide insight into the unknown of functional diversity. By isolating individual subsets of the community, and monitoring their nutritional strategies and performance under stress, we can gain valuable information into responses from different management systems.

Keywords. Functional diversity, microbial communities, substrate utilization, tillage, landscape

1. Introduction

Anthropogenic influences can affect ecosystem functioning and diversity, with some of the most dramatic examples of ecosystem disturbances occurring as a result of soil erosion. Microorganisms impact agroecosystems in a variety of ways. Fluxes in microbial diversity with disturbance may contribute greatly to the understanding of functional diversity as it relates to soil quality and the development of sustainable agroecosystems (Hawksworth, 1991; di Castri and Younes, 1990; Thomas and Kevan, 1993). Diversity measures are evolving from a quantitative measure of richness and evenness, such as the Shannon-Weaver Index, to more qualitative measures involving substrate utilization analyses (Garland, 1996b) and fatty acid methyl ester (FAME) analyses (Zelles et al., 1995). While rates of microbial activity are based on quantitative measures, microbial functions are associated with qualitative measures identifying specific groups of the community. Soil microorganisms are one of the most sensitive biological markers available and the most useful for classifying disturbed or contaminated systems since diversity can be affected by minute changes in the ecosystem. The use of microbial functioning for examination of environmental stress and changes in biological diversity needs to be exploited for the benefit of soil quality within agroecosystems (OTA, 1987; Kennedy and Papendick, 1995).

Substrate-use patterns based on sole carbon source utilization groupings can characterize bacterial communities (Zak et al., 1994). Viable organisms will respond to the substrates that they can metabolize, providing an overall community fingerprint. Substrate utilization measures of diversity provide the basis for grouping soils with respect to microbial community similarities and differences. Shifts in substrate utilization measures of diversity may mirror changes in microorganisms performing similar functions. These changes also can be viewed as differences in the availability of carbon substrate as dependent on the composition of root exudates from different plant species (Garland, 1996b). Individual communities can be separated through principle component analysis of the unique combination of substrates that the community can utilize (Zak et al., 1994; Garland, 1996a). The usefulness of substrate utilization is dependent on its ability to consistently characterize microbial communities based on sole carbon source utilization and to elucidate changes in community response due to some perturbation (Garland,1996b).

Recent studies (Zak et al., 1994; Bossio and Scow, 1995; Garland, 1996ab) have focused on the use of Biolog® plates to obtain information as to the metabolic diversity of microbes of soil systems. These studies have been successful in assessing differences due to moisture and plant communities; however, the overall function of the microbial community still needs to be addressed. Zak et al. (1994) divided the carbon sources into guilds, or classes, such as carbohydrates, amino acids, polymers, etc. and assessed the differences among the microbes associated with six different plant communities. Substrate use as a biomarker for carbon availability in the rhizosphere simplifies the monitoring of organic materials released by the roots. Substrate utilization also has been used to monitor carbon source utilization based on the assumption that shifts in utilization are linked to availability in the environment (Root, 1967; Garland, 1996b). This method also may provide a faster, easier measure of community diversity than phospholipid fatty acids (Garland, 1996b).

Reduction in aboveground plant diversity that occurs with severe disturbance such as tillage, overgrazing, and pollutants decreases microbial diversity as based on the Shannon index (Gochenauer, 1981; Boddy et al., 1988; Christensen, 1989). In a study of the diversity of prairie and cultivated soils, diversity indices as determined by substrate utilization were greater in disturbed or cultivated systems when compared to grassland (Table 1; Kennedy and Smith, 1995). While the traditional diversity indices may provide insight into community shifts, substrate utilization may indicate changes in the microbial community to one of greater substrate utilization diversity and stress resistance (Kennedy and Smith, 1995).

Our objective was to characterize soil community structural differences with respect to type of management through analyses of substrate utilization on both bacterial isolates and whole soil samples. By applying substrate utilization measures of diversity, we can identify similarities of community structures with management, and increase our understanding of the effect of the community structure on functional diversity.

2. Materials and Methods

Plow layer soil (0 to 30 cm) was collected from two sites in the Pacific Northwest, near Pullman WA, USA. The sites had natural prairie and cultivated land plots on same slope and aspect in close proximity and were the same sites used in a previous study (Kennedy and Smith, 1995). Samples were collected from the two sites in a grid pattern across the toposequence. The cultivated site was in a wheat-barley-pea rotation and the prairie had not been disturbed since the family had homesteaded 100 years ago. The soil series was a Palouse silt loam. Soil was collected in layers to a depth of 30 cm and refrigerated until analysis. All analyses were conducted within four weeks of sampling. Any plant patches were avoided in the prairie and the interrow areas were sampled in the cultivated system. Nine samples were collected at each of three toposequences and analyzed as individual samples.

We examined heterotrophic bacterial communities in the soil. Soil was plated on various media selected to represent broad nutritional groups representing the major microbial groups found in soil. The media used were tryptic soy agar (bacteria), Sands and Rovira medium (Gram⁻ bacteria), and starch casein (actinomycetes; Wollum, 1982). Plate counts were taken and morphological differences noted. To obtain isolates, 60 isolates for each sample treatment were randomly selected and streaked from plates containing the highest number of separate colonies. These isolates were cultured in broth of their initial medium and stored in that medium plus 40% glycerol at -80°C. The characteristics of each isolate were determined to calculate diversity indices (Simpson, 1949; Shannon and Weaver, 1949; Hill, 1973). Each isolate was analyzed for its ability to utilize substrates representing various functional groups and their ability to withstand stress. Starch hydrolysis, proteolytic activity, lypolysis, growth at pH 4, 7, and 9 and ability to grow in the presence of heavy metals, polyethylene glycol, or antibiotics (Gerhardt, 1981) were tested. Substrate utilization was determined using a basal growth medium containing several substrates including glutamate, glycine, citrate, serine, xylan, or other hemicelluloses, cellulose, and mannans as the sole source. All parameters were screened using 96-well microtiter plates, and growth was indicated by the redox dye, triphenyl tetrazolium chloride (Kennedy, 1994). Seven selected carbon sources were analyzed for percent utilization for isolates of both wheat and grassland systems. The sources included simple sugars, ketones, and complex lignins. The percentage of each source that an individual sample or group of samples could utilize was recorded. The ten isolates were taken as replications for each landscape position. Tallies of the categorical data were compiled under collapsed versions of the data set. Associations between the variables and the substrates were calculated using a logit model chi-square test. The values are reported as a Ward statistic (Agresti, 1990).

3. Results and Discussion

3.1 Isolates analyses

In a previous study of the diversity of prairie and cultivated soils, diversity indices were generally greater in disturbed or cultivated systems when compared to prairie grassland (Table 1; Kennedy and Smith, 1995). The ecology of root-microbe interactions in minor disturbance is different from that after extensive plowing to prepare the seedbed. Changes in a soil's physical and chemical properties resulting from tillage greatly alter the matrix supporting growth of the microbial population. In other studies of no-till agricultural systems, microbial activities differed with depth, with the greatest microbial activity occurring near the no-till surface; in the tilled system, activities were more evenly distributed throughout the plow layer (Doran, 1980).

Table 1. Richness, evenness, and diversity values for microbial populations from soil from a grass prairie (p) and conventionally cultivated wheat field (ct) in eastern Washington (Kennedy and Smith, 1995)

	Prairie	Conventional Till	
Value	Index	Index	Relationship
Richness			
Margalef	31.18	37.58	p<ct
Menhinick	17.33	19.42	p<ct
Evenness			
Hill mod.	0.446	0.165	p>ct
Diversity			
Shannon	1.05	1.89	p<ct
Hill	2.86	6.67	p<ct

Soil microbial diversity was assessed for heterotrophic bacterial communities in the soil from both sites using substrate utilization patterns and resistance to antibiotics, metals or polyethylene glycol. The rankings, thus the characteristics, of each species unit were used to calculate diversity indices (Simpson, 1949; Shannon and Weaver, 1949; Hill, 1973). The increase in diversity with disturbance indicates a change in the bacterial community to one that exhibited a greater range of substrate utilization and stress resistance. Upon further investigation, the populations were analyzed for substrate utilization with regard to management and landscape position. Tallies of the number of isolates utilizing the substrates under collapsed versions of the data provided patterns of substrate usage according to a particular variable. Diversity indices can be used to indicate the effect of disturbance; however, diversity numbers may not indicate the full diversity within a system. Greater or lesser diversity should not be equated with a system, rather the changes in diversity with management may be more informative of the status of a soil microbial community.

While an assessment of the overall microbial carbon source utilization revealed little difference between management types (Fig. 1), differences were seen when these data were analyzed based on type of isolates utilizing the carbon. The pattern for bacterial and actinomycetes isolates differed between these systems. The systems initially were similar with regard to substrate, but the functional groups from each system were different. Prairie systems relied on the fungal population for decomposition while in conventionally-tilled systems, the bacterial component was found to be responsible for a greater portion of residue decomposition. These studies illustrate the alteration of the makeup of the microbial communities and possibly the diversity of basic microbial groups with changes in management systems.

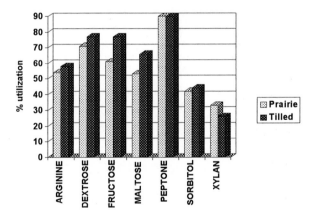

Fig. 1. A comparison of overall microbial carbon source utilization patterns from bacteria isolated from prairie and tilled soils.

We also found changes in bacterial composition of this system. The till bacterial isolates utilized more of the carbon sources, except peptone, than the prairie bacterial isolates (Figure 2a), while the prairie actinomycetes could utilize more of the carbon sources than the till actinomycetes (Figure 2b). Overall, actinomycetes populations from both systems had a higher percentage of utilization for all of the carbon sources than the bacteria.

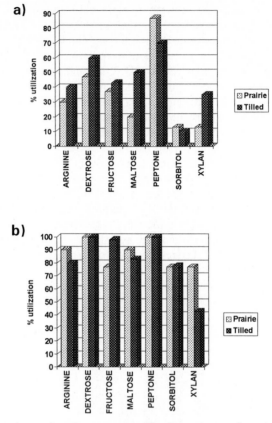

Fig. 2. A comparison of carbon source utilization patterns from bacteria (a) and actinomycetes (b) from prairie and tilled soils.

The effects of landscape position on substrate utilization were also analyzed (Fig. 3). The percentage of carbon utilization was affected by origin on the landscape. Samples of isolates were divided into foot slope (Fig. 3a), side slope (Fig. 3b) and ridge top (Fig. 3c) positions and analyzed for carbon utilization. The overall percent utilization from all sources in the tilled system were highest at the foot slope which would be expected due to higher organic matter content at this landscape position. There was a trend in the wheat system towards higher utilization of the more simple sugars in the side slope and ridge top. The overall percent utilization from all sources in the grassland system was highest at the

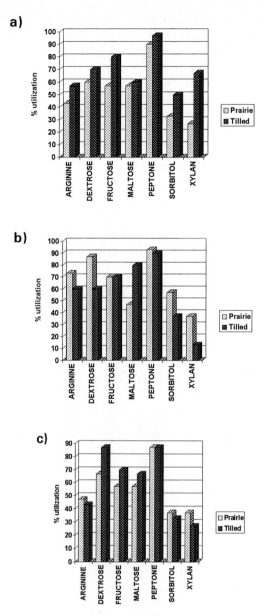

Fig. 3. A comparison of carbon source utilization of prairie and tilled soils for foot slope(a), side slope(b) and ridge top (c).

side slope position. When comparing the two systems based on total percent utilization at the individual landscape positions, tilled isolates utilized more of all of the carbon sources at the foot slope, prairie isolates utilized more of the carbon sources at the side slope, and both had similar patterns at the ridge top. These differences in substrate utilization further illustrate the use of patterns to distinguish changes in microbial populations and functioning with management changes.

The use of stressors to monitor isolate response also provides information about the success of different groups of organisms to survive environmental changes such as potential desiccation or high metal content. Actinomycetes isolates from both tilled and prairie systems had the highest response to such stresses (Fig. 4c). No major differences were seen in stress response in either the tilled or prairie system.

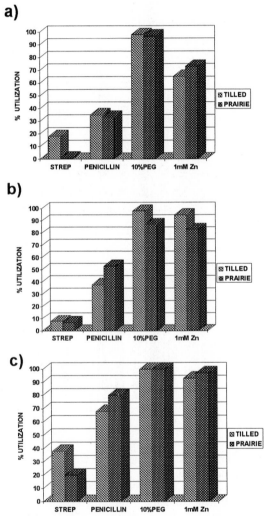

Fig. 4. Percent utilization of a) bacteria, b) fungi, and c) actinomycetes isolates in wheat and grass systems in response to stress.

3.2 Relationship analyses

Logit models are conservative measures of substrate utilization data which can reveal associations between data and variables. Among the variables in this study were isolate type, management system, and landscape position. We found relationships among isolate type and management system (Table 2). Arginine, sorbitol, and xylan were substrates found to have significant relationships among the isolates. Actinomycetes were better able to utilize these sources than either bacteria or fungi. Actinomycetes were also better able to respond to stressors, specifically the presence of streptomycin and penicillin. The substrates analyzed may provide insight into the type of substrates available in these systems as well as the relative complexity of those substrates. More isolates from the tilled system utilized maltose than the isolates from the prairie system, while more prairie isolates utilized xylan. Simple sugars may be more prevalent in tilled systems, while the more complex substrates, such as xylan, may be more prevalent in undisturbed systems such as prairie. The availability of these substrates may be due to either the plant community, root exudates or microbial exudates. No relationships were identified for dextrose, fructose, peptone, polyethylene glycol or zinc.

Table 2. Chi-square values and relationships among isolate type and management for substrate utilization data.

Substrate	x^2 for isolate*	x^2 for system	Relationship
Carbon source			
Arginine	38.71	NS	actinomycetes>fungi>bacteria
Dextrose	NS	NS	
Fructose	NS	NS	
Maltose	NS	4.60	tilled > prairie
Peptone	NS	NS	
Sorbitol	44.76	NS	actinomycetes>fungi>bacteria
Xylan	47.09	4.33	actinomycetes>bacteria>fungi; Prairie > tilled
Stressor			
Streptomycin	19.67	NS	actinomycetes>fungi>bacteria
Penicillin	36.27	NS	actinomycetes>bacteria>fungi
Polyethylene glycol	NS	NS	
Zinc	NS	NS	

*Reported as Ward statistic (Agresti, 1990); NS = not significant

4. Conclusions

Microorganisms are key to agroecosystem functioning by their involvement in nutrient cycles and decomposition, soil structure, and plant growth. We need to increase our understanding of the diversity and function of microbial communities in agroecosystems. Substrate utilization is one method to achieve this, especially due to the wealth of information that can be quickly obtained. Patterns of substrate utilization vary with landscape position and management. These patterns are also different for each type of microbe in the system. These data illustrate that the use of diversity measurements to assess management impacts may be limited. The use of functional groups can give us further information on the status of a system; however, this type of analysis needs further study. The challenge ahead is to identify the level of microbial diversity, species composition, distribution, and the resiliency of the community to withstand stress and maintain an ecosystem. We need to determine the extent of microbial diversity, both taxonomic and functional, in agroecosystems and increase our knowledge of the functional roles of microbes to assess their role in agroecosystem quality and productivity.

5. Acknowledgements

Support from the U. S. D. A. - Agricultural Research Service in cooperation with the College of Agriculture and Home Economics, Washington State University, Pullman, WA 99164 is appreciated. Trade names and company names are included for the benefit of the reader and do not imply endorsement or preferential treatment of the product by the U. S. Department of Agriculture or Washington State University. All programs and services of the U.S. Department of Agriculture are offered on a nondiscriminatory basis without regard to race, color, national origin, religion, sex, age, marital status, or handicap.

6. References

Agresti A (1990) Categorical Data Analysis. John Wiley & Sons, New York
Boddy L, Watling R, Lyon AJE (1988) Fungi and ecological disturbance. Proc R Soc Edinb Sect B Vol 94 1988
Bossio DA, Scow KM (1995) Impact of carbon and flooding on the metabolic diversity of microbial communities in soils. Appl Environ Microbiol 61: 4043-4050
Christensen M (1989) A view of fungal ecology. Mycologia 81:1-19
di Castri F, Younes T (1990) Ecosystem function of biological diversity. Biol Int Special Issue 22:1-20
Doran JW (1980) Soil microbial and biochemical changes associated with reduced tillage. Soil Sci Soc Am J 44: 765-771
Garland JL (1996a) Analytical approaches to the characterization of samples of microbial communities using patterns of potential C source utilization. Soil Biol Biochem 28: 213-221
Garland JL (1996b) Patterns of potential C source utilization by rhizosphere communities. Soil Biol Biochem 28: 223-230

Gerhardt P (1981) Manual of Methods for General Bacteriology. American Society for Microbiology. Washington, DC. 524 p

Gochenauer SE (1981) Responses of soil faunal communities to disturbance. In: Wicklow DT, Carroll GC (eds) The Fungal Community: Its Organization and Role in the Ecosystem, Marcel Dekker, Inc., New York, pp 459-479

Hawksworth DL (1991) The Biodiversity of Microorganisms and Invertebrates: Its Role in Sustainable Agriculture. CAB International, Redwood Press Ltd., Melksham, UK. 302p

Hill MO (1973) Diversity and evenness: A unifying notation and its consequences. Ecology 54: 427-432

Kennedy AC (1994) Carbon utilization and fatty acid profiles for characterization of bacteria. In: Weaver R and Angle JS (eds) Methods of Soil Analysis, Part 2. Microbiological and Biochemical Properties, Soil Science Society of America. Madison, WI, pp 543-554

Kennedy AC, Papendick RI (1995) Microbial characteristics of soil quality. J Soil Water Conserv 50: 243-248

Kennedy AC, Smith KL (1995) Soil microbial diversity and ecosystem functioning. Plant Soil 170: 75-86

Office of Technology Assessment. U. S. Congress (1987) Technologies to Maintain Biological Diversity. OTA-F330, U.S. Gov. Print. Office, Washington D.C. 334 p

Root RB (1967) The niche exploitation pattern of the blue-gray gnatcatcher. Ecological Monographs 37: 317-350

Shannon CE, Weaver W (1949) The Mathematical Theory of Communication. University Illinois Press, Urbana, IL. 117 p

Simpson EH (1949) Measurement of diversity. Nature 163: 688

Thomas VG, Kevan PG (1993) Basic principles of agroecology and sustainable agriculture. J Agric Environ Ethics 5:1-18

Wollum II, AG (1982) Cultural methods for soil microorganisms. In: Page AL, Miller RH, Keeney DR (eds) Methods of Soil Analysis Part 2 Chemical and Microbiological Properties, American Society of Agronomy, Madison, WI, pp 781-802

Zak JC, Willig MR, Moorhead DL, Wildman HG (1994) Functional diversity of microbial communities: A quantitative approach. Soil Biol Biochem 26: 1101-1108

Zelles L, Bai QY, Rackwitz R, Chadwick D, Beese F (1995) Determination of phospholipid-derived and lipolysaccharide-derived fatty acids as an estimate of microbial biomass and community structures in soils. Biol Fert Soils 19: 115-123

Different Carbon Source Utilization Profiles of Four Tropical Soils from Ethiopia

Shobha Sharma[1], Alessandro Piccolo[2] and Heribert Insam[1]

[1]Institut für Mikrobiologie, Universität Innsbruck, Technikerstr. 25, A-6020 Innsbruck, Austria.
[2]Department of Agricultural Chemical-Science, University of Naples, Napoli, Italy

Abstract. Microorganisms play a crucial role for maintaining the fertility and productivity of soils. Understanding the functional abilities of microorganisms may help to improve agricultural practices. While the Biolog approach is increasingly being used for characterising the soil microbiota in temperate soils, it has rarely been used for tropical soils. In this study we investigated four tropical soils from different climatic regions of Ethiopia (Jima, Awassa, Holeta and Ginchi) receiving different types of organic fertilizers. With discriminate analysis we were able to differentiate between the sites. When all sites were computed together, the functional diversity of the microbial community was significantly correlated with the microbial biomass (C_{mic}). By multivariate analysis of variance (MANOVA), we were able to identify the substrates which were significant for the separation between the sites. The results encourage us to further use the substrate test for investigating fertilizer effects on the microbial communities of tropical soils.

Key words. Organic amendments, tropical soil, substrate utilization, Biolog, Africa, fertilizer

1. Introduction

In Ethiopia, agriculture provides employment for 86 percent of the labour force (Ayele and Mamo, 1995). To support the Ethiopian farming population, soil management practices must be oriented towards a sustainable crop production. In this regard, advanced knowledge on the status of soil organic matter is fundamental to establish a sustainable soil management. Microbial biomass constitutes the active fraction of soil organic matter and the fertilization of soil and cropping practices affect microbial growth and activity (Goyal et al. 1992). The productivity of agricultural systems is also greatly influenced by the functional abilities of soil microbial communities (Pankhurst et al. 1996). The Biolog approach has been increasingly used for characterising soils from different agricultural ecosystems in temperate climates. However, few studies have been conducted for tropical soils. In this study we investigated four tropical soils from

Ethiopia that received different organic amendments for improving their carbon status. The aim was to detect if the soil management changes the functional abilities of the soil microbial community.

2. Materials and Methods

2.1 Site description

Soils from four different locations in southern Ethiopia were used for the study. Jima is situated at an altitude of 1800m and receives an average rainfall of 1400 mm per year. The mean annual temperature (MAT) is 18.4°C. Coffee production intercropped with maize and other local cereal is commonly practiced. The present trial has been done since 1992. Awassa is located at an altitude of 1700m and receives an average rainfall of 1100 mm per year. The MAT is 19.5°C. The field trials which began in 1992 involve the incorporation of coffee residues along with urea into the soil. Holeta is situated at an altitude of 2400m and receives an average rainfall of 1050 mm per year. The MAT is 22°C. The field trials investigate the effect of mustard meal, incorporated into the soil. The trial was started in 1992. Crops cultivated are wheat and barley. Ginchi is located at an altitude of 2300m. The mean annual rainfall and the MAT are 1300 mm and 16.2°C respectively.

2.2 Soils and Sampling

The soils were sampled from the top 15 cm of the A horizon (the main soil physical and chemical properties are given in Table 1). Before analysis the soils were passed through 2 mm mesh size sieve and were stored at room temperature in CO_2 permeable polyethylene bags. The description of the treatment is given in Table 2.

2.3 Extraction and microtiter plate preparation

To obtain the bacterial suspension, 10 g soil was mixed with 90 ml of Ringer solution (quarter strength) and was shaken for 1h. The suspension was allowed to settle for 15 minutes and the supernatant was used for further dilution. Three replicate microtiter plates were inoculated with the final dilution which contained approximately $4.0*10^4$ cells per well (acridine orange direct count, AODC) and were then incubated at 30°C. After the appearance of colour in some wells, the plates were read every 6 hours up to 3 days at 592 nm with a microtiter plate reader (SLT SPECTRA, Grödig, Austria).

2.4 Microbial biomass and total organic carbon

Microbial biomass C_{mic} was determined by substrate induced respiration (SIR) (Anderson and Domsch, 1978) with a continuous flow analyser (Heinemeyer *et al.*, 1989). Total organic carbon was determined by dry combustion with a carbon analyser (Ströhlein Instruments C-Mat 550 PC).

Table 1. Main soil physical and chemical parameters

Sites	Soil type	pH (CaCl$_2$)	Texture [%]			Organic Carbon (C$_{org}$) [%]
			Sand	Silt	Clay	
Jima	Eutric Nitosol	4.72	-	16.25	83.75	2.02
Awassa	Dystric Cambisol	6.08	42.50	30.00	27.50	1.66
Holeta	Chromic Vertisol	4.61	-	16.25	83.75	1.60
Ginchi	Vertisol	6.44	2.50	26.25	71.25	2.00

Table 2. Different soil treatments

Site	Soil Treatment
Jima	Virgin (forest soil) Control (cultivated soil, no amendment) Coffee husk (3-9 tons/ha per year) Maize residue Maize residue + (3-9 tons/ha coffee husk per year) Cowdung Cowdung + (3-6 tons/ha coffee husk per year)
Awassa	Virgin (forest soil) Control (cultivated soil, no amendment) Urea Urea + (10-20 tons/ha coffee husk per year) Coffee husk (10-20 tons/ha coffee husk per year)
Holeta and Ginchi	Virgin (forest soil) Control (cultivated soil, no amendment) Mustard meal (90-180 kg/ha per year)

2.5 Data analysis and statistics

For the further analysis of Biolog data a Riemann's sum for each substrate was calculated from all the readings taken during 3 days of incubation. The Riemann's sum may be regarded as an approximation of the area under the curve. It can be expressed as follows:

For substrate i,

$$S_i = 1/T[(t_1-t_0) (x_{i,1} + x_{i,0})/2 + (t_2-t_1) (x_{i,2} + x_{i,1})/2 + ... + (t_j-t_{j-1}) (x_{i,j} + x_{i,j-1})/2]$$

where
T=t_j-t_0 (t_0 is the incubation hour when first reading was made
   t_j is the incubation hour when *j* reading was made)
$x_{i,j}$= absorbance value of *i* substrate measured at the point t_j

Stepwise discriminant analysis was applied to the Biolog data set to identify the substrates that contributed most for the discrimination among the communities from the four sites ($p \leq 0.05$ for a substrate to enter or remain in the function). The substrates selected by stepwise procedure were further subjected to multivariate analysis (MANOVA) to determine which of these were significant for the differentiation among communities from each site (Bonferroni confidence interval, $p \leq 0.05$). For the statistical analyses the SPSS statistical package (SPSS, 1994) and Statistica program package (StatSoft, 1995) were used.

2.6 Diversity

The functional diversity of the microbial community was calculated using the Shannon diversity index according to Zak *et al.* (1994):
$$H' = - \sum p_i (\log p_i)$$
where p_i is the ratio of the absorbance value of each well to the sum of absorbance values of all wells .

3. Results

3.1 Discriminant analysis

The discriminant analysis clearly separated the soils from different sites (Fig. 1A). The vertisols (receiving similar types of treatment) from Holeta and Ginchi appeared close to each other, indicating the similarity of the microbial community in these similar soil types as compared with the other soils. To get further information on the effect of different soil treatments, discriminant analyses were separately performed for soils from each site. In the case of Jima the communities from the soils receiving different organic fertilizers were well separated on the axis of function 1 (Fig. 1B). The discriminant function 2 separated only the forest soil from the other treatment types. Soils treated with coffee husk and maize residue mixed with coffee husk appeared close to the control. The soils receiving manure (cowdung) were separated from other types on the axis of function 1. The effect of organic matter input could be also perceived for soils from Awassa (Fig. 1C). On the axis of function 1 only soils receiving urea were separated from the others. However, on the axis of function 2 other three types, i.e., soils treated with coffee husk, control and the forest soil were well differentiated. Similarly the treatment effects were found in the soils from Holeta and Ginchi (Fig. 1D). In this case, however, only the virgin soils of both sites were close to each other.

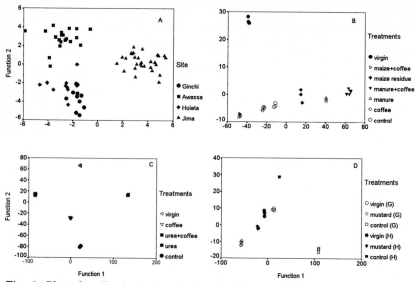

Fig. 1. Plot after discriminant analysis of soils from different sites **(A)**, plots after discriminant analyses from the soils receiving different organic inputs from Jima **(B)** Awassa **(C)** Holeta (H) and Ginchi (G) **(D)**

3.2 Individual substrates

The stepwise discriminant analysis selected 15 substrates most important for the separation among different sites. However, only two of these substrates were significant for the separation of Ginchi from Holeta, whereas 9-11 substrates were significant for the separation of Holeta and Ginchi from Awassa and Jima. Likewise, the communities from Awassa and Jima were significantly differentiated by 6 substrates (Table 3). This indicated that the microbial activities of the communities from Holeta and Ginchi, separated only by 2 substrates were similar to each other.

3.3 Diversity

The functional diversity (Shannon diversity index, H') of the bacterial communities from Awassa and Jima was significantly higher than that of the Holeta and Ginchi communities. No clear relationship between soil treatment and diversity was found. Only in the case of soils from Holeta and Ginchi the incorporation of mustard meal has increased the functional diversity of microbial communities (Table 4).

3.4 Microbial biomass (C_{mic}), organic carbon (C_{org}) and C_{mic}/C_{org} ratio

Microbial biomass, organic carbon and C_{mic} / C_{org} ratio of the soils from Awassa were significantly higher compared to other sites. C_{mic} and C_{org} of soils from Jima were also significantly different from that of Holeta and Ginchi. However no significant differences were found between the soils from Holeta and

Table 3. Substrates which were significantly different among the indicated sites

Substrates	Awassa-Holeta	Awassa-Ginchi	Jima-Holeta	Jima-Ginchi	Jima-Awassa	Holeta-Ginchi
Cellobiose	X	X			X	
L-fucose	X	X	X	X	X	x
α-D-glucose	X	X	X	X		
m-inositol	X	X	X	X		
Lactulose	X		X	X		
D-trehalose			X	X		x
Quinic acid	X		X			
Sebacic acid			X	X	X	
Bromo succinic acid	X					
L-phenylalanine		X	X	X	X	
Urocanic acid	X	X		X	X	
Inosine	X	X		X		
Uridine	X	X			X	
Phenyl ethylamine			X			
Glucose-6-phosphate	X	X				

Table 4. Functional diversity, microbial biomass and total organic carbon from soil receiving different organic inputs from all sites

Site	Soil Treatment	diversity index [H']	C_{mic} ($\mu g\ C_{mic}\ g^{-1}$ dry soil)	C_{org} (%)	C_{mic}/C_{org} (mg $C_{mic}\ g^{-1}\ C_{org}$)
Jima	virgin (forest soil)	1.86	216.57	3.36	6.43
	control	1.83	128.83	2.17	5.94
	coffee husk	1.79	111.05	2.18	5.08
	maize residue	1.85	115.86	2.27	5.09
	maize residue+coffee husk	1.78	113.23	2.09	5.40
	cowdung	1.80	106.53	2.35	4.54
	cowdung +coffee husk	1.81	98.22	1.85	5.30
Awassa	virgin (forest soil)	1.83	-	2.82	-
	control	1.80	163.77	1.51	10.82
	urea	1.87	140.11	1.40	10.06
	urea+coffee husk	1.83	152.90	1.88	8.13
	coffee husk	1.84	211.72	1.99	10.70
Holeta	virgin (forest soil)	1.50	177.26	2.51	7.05
	control	1.71	79.82	1.33	5.55
	mustard meal	1.81	101.2	1.12	7.42
Ginchi	virgin (forest soil)	1.67	138.21	1.68	8.22
	control	1.68	97.52	1.37	7.09
	mustard meal	1.75	126.25	1.38	9.13

- ... not determined

Ginchi. For all sites C_{mic} was significantly correlated with C_{org} ($r = 0.503$, $P < 0.001$)

When all sites were computed together C_{mic} was significantly correlated with the functional diversity of the microbial communities (Fig. 2). However, no significant correlation was found for the soils from Awassa and Holeta when correlation was done for the individual sites.

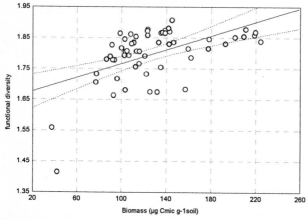

Fig. 2. Correlation between microbial biomass C and the functional diversity for all sites ($r = 0.536$, $P = 0.001$)

4. Discussion

In agroecosystems large amounts of biomass and nutrients are removed as crops. In the tropics, the control of soil organic matter and nutrient content may be achieved by the addition of organic inputs. The degradation rate of this added organic material and the composition and distribution of the degradation products to soil particles affect the stability of the native organic matter pool. Since soil microbial communities play a significant role in the degradation process, the study of the functional abilities of the microbial communities is crucial for evaluating the quality and the effects of such organic inputs.

The Biolog approach has been successfully used to characterize microbial communities from different ecosystems (Garland and Mills 1991, Winding 1994, Zak et al. 1994, Insam et al. 1996). A decrease in functional diversity of microbial communities during maize litter decomposition was observed by Sharma et al. (submitted).

The results of this study showed that the use of the Biolog approach could be valuable to examine the effects of organic inputs on the microbial communities from tropical soils. The effects of treatments on C_{mic} and C_{org} were not pronounced. However, the variations among communities from soils receiving different fertilizers were clearly detected by the Biolog approach. Further, by MANOVA we were able to detect substrates which were significant for the separation between sites (Table 4). Though no clear relationship between soil treatment and Shannon diversity index (H') was observed, we found a significant correlation between H' and microbial biomass. This indicates that an increased microbial biomass is not caused by the enhancement of only few distinct bacterial groups. This may be an important advantage of organic amendments over chemical fertilization.

5. Acknowledgements

This study was supported by grants from the European Union (No. ERB 3504 PL921207) and the Austrian Science Fund (Project No. P 10186-GE0).

6. References

Anderson JPE, Domsch KH (1978) A physiological method for the qualitative measurement of microbial biomass in soils. Soil Biol Biochem 10: 215-221

Ayele G, Mamo T (1995) Determinants of demand for fertilizer in Vertisol cropping system in Ethiopia. Trop Agric 72: 165-169

Garland JL, Mills AL (1991) Classification and characterization of heterotrophic microbial communities on the basis of patterns of community-level sole-carbon-source utilization. Appl Environ Microbiol 57: 2351-2359

Goyal S, Mishra MM, Hooda IS, Singh R (1992) Organic matter-microbial biomass relationship in field experiments under tropical conditions: effects of inorganic fertilization and organic amendments. Soil Biol Biochem 24: 1081-1084

Heinemeyer O, Insam H, Kaiser E, Walenzik G (1989) Soil microbial biomass measurements: an automated technique based on infra-red gas analysis. Plant Soil 166: 191-195

Insam H, Rangger A, Henrich M, Hitzl W (1996) The effect of grazing on soil microbial biomass and community on alpine pastures. Phyton 36: 205-216

Pankhurst CE, Ophel-Keller K, Doube BM, Gupta VVS (1996) Biodiversity of soil microbial communities in agricutural systems. Biodiversity and Conservation 5: 197-209

Sharma S, Rangger A, Insam H (submitted) Change in functional diversity of microbial communities during maize litter decomposition

SPSS (1994) Statistical package of the soil sciences. SPSS Inc., Chicago.

StatSoft, Inc. (1995) Statistica for windows [computer program manual]

Winding A (1993) Fingerprinting bacterial soil communities using Biolog microtiterplates. In: Ritz K, Dighton J, Giller KE (eds) Beyond the Biomass Wiley & Sons, London, pp. 85-94

Zak JC, Willig MR, Moorhead DL, Wildman HG (1994) Functional diversity of microbial communities: a quantitative approach. Soil Biol Biochem 26: 1101-1108

Fallow Age Influences Microbial Functional Abilities, Soil Properties and Plant Functional Groups

Shivcharn S. Dhillion

Department of Biology and Nature Conservation, Agricultural University of Norway (NLH), PB 5014, Ås, N-1432, Norway

Abstract. In West Africa local intensification has caused the shortening of fallow periods and uncontrolled land use/abandonment, contributing to gradual land degradation. Recently, to combat abandonment, the revival of the traditional cultural practice of fallow has been encouraged, and interest in understanding the processes associated with fallow is growing. Soil from fallow fields (ages 1, 3, 7, 14) was used to evaluate the relationships of several biotic and abiotic variables with the age of fallow to shed light on the usefulness of fallow in potentially building up ecosystem properties. In addition a microbe substrate utilization approach was used to assess the functional diversity of soil microorganisms associated with soil of agricultural and fallow communities. Specifically the following question is addressed: What are the relationships of the microbial functional abilities, soil properties and plant functional groups on former agricultural land? The age of fallow was positively correlated with the number of microbial functional activities. There were no significant differences between the cropped sites and the one year old sites for all the functional diversity indexes (richness, evenness, diversity), whereas soil from the 3 year old fallow sites was significantly different from both the younger and older fallows, falling in-between the two groups. Although substrate richness was significantly higher in soils from the 7 year fallow than those of the 14 year fallow, substrate diversity and evenness were higher in the 14 year fallow. The canonical discriminate analysis was able to separate the sites by fallow age with two functions. Certain groups of substrates may be significant to the understanding the changes occurring in the soil, for example the amino acids separated sites well. Furthermore, several inorganic soil nutrients, organic matter, microbial biomass carbon, hyphal lengths, number of mycorrhizal spores, enzymatic activity, and the density of perennial (herbaceous and woody) species had positive relationships with the increase in age of fallow. All relationships suggest that there may be differences in resources available along the chronosequence of fallow. The data supports the view that several soil properties, biotic in particular, are enhanced with fallow age via several possible mechanisms mediated by the establishment of perennial species, and the interactions of inputs, presumably via roots, with soil biota and nutrients. Therefore the practice of fallow ought to be encouraged, however precise management recommendations will require longer term studies. This method, by providing rapid initial insight to the functional abilities of the microbes, may be seen as a tool to point towards the potential areas of exploration where more sophisticated methods can be utilized. This study also provides evidence that the simultaneous study of soil nutrients, vegetation composition and inputs, and soil microbial activity is essential in understanding processes associated with land use practices like fallow.

Keywords. Functional groups, land use, diversity, soil enzymes, fallow, BIOLOG, vegetation, abandonment

1. Introduction

Global trends in agricultural intensification, and the need to conserve and restore biodiversity, have created unpredictable futures, especially for marginal lands (Heywood and Watson, 1995; Ricklefs and Schluter, 1993; Floret and Pontanier, 1993). In West Africa (especially in the subtropical and tropical zones) land is often abandoned or set-aside as fallow when productivity is very low in the hope that it would recover for reuse. Local intensification has however caused the shortening of fallow and uncontrolled usage of land. Such practices lead to subsequent abandonment of land, contributing to gradual land degradation. Recently consideration has been given to encourage the revival of the traditional cultural practice of fallow, and interest in understanding the process and function of fallow is growing. For villagers in West-Africa, fallow can be economically important, for example used for grazing, and collecting tubers, wood, and medicinal/spice plants (Floret and Pontanier, 1993).

The activities and interactions of soil microbial communities with soil flora/fauna and plant roots have only recently been highlighted as one of the most important for inclusion in community and ecosystem studies. An integral part of understanding their role in the system as a whole is exploring their relationships with both biotic and abiotic components. While their activities are likely influenced by environmental changes and land use practices our knowledge of microbes is limited (Atlas, 1984; Parkinson and Coleman, 1991; Hooper et al., 1995). Costly and time consuming methodology is perhaps the biggest constrain in terms of studying microorganisms.

A global assessment of microbial and ecosystem processes has highlighted the general lack of information relating diversity variables to aspects of ecosystem function (Hooper et al., 1995; see also within Heywood and Watson, 1995). Recent studies show that in plant communities, diversity increases resistance to disturbances, soil mineral nitrogen usage, CO_2 flux, productivity, and accumulation of inorganic minerals (see details in Mooney et al., 1995). Elevated CO_2 levels can also impact directly or indirectly root growth (substrate input), mycorrhizal associations, soil microbial biomass carbon, and enzyme activities in soils of increasing plant diversity ecosystems (Roy et al. 1995; Dhillion et al. 1996). These studies suggest that plant community diversity and/or plant compositional differences may introduce a range of resources (substrates) which in turn could influence the dynamics of microbes in soil.

In this study the relationships of soil variables (biotic and abiotic) with the age of fallow were evaluated to explore the usefulness of fallow as a land use practice in potentially building up ecosystem properties. A new microbe substrate utilization approach, Biolog, was used to assess the functional diversity of soil microorganisms of agricultural and fallow communities in Mali.

Specifically the following question is addressed: What are the relationships of microbial functional abilities and soil properties on former agricultural land?

2. Methods

2.1 Site

Fallow is a common cultural practice of many ethnic groups in Mali (Floret and Pontanier, 1993). In general, fallow periods averaged about 5-7 years. However due to agricultural intensification and population growth, the fallow periods have shortened to about 3 years (Floret and Pontanier, 1993; personal observation). Despite the shortening of fallow periods, villagers still believe that the age of the fallow causes increase in quality of the land, allows establishment of valuable tree species and medicinal plants, and brings peaceful spirits into the region (Floret and Pontanier, 1993). The latter two believes allow some areas to be designated for longer periods. As part of a land rehabilitation project, sampling was possible on a range of sites in the Siccasso area in the subtropical region of Mali. Using a stratified random sampling procedure, herbaceous vegetation sampling was done by placing quadrates (1x1m) along transects within sites. Entire sites were sampled for woody species densities. All sites chosen had similar soil type, slope and elevation. Details on these aspects are presented in another study (Dhillion, in preparation). The last crop planted on the fallow sites was sorghum, as recalled by the village elders. Sites were comprised of:

Agricultural fields (n=2) of sorghum not differing in cropping history and separated by a buffer zone of trees.

Fallow fields available were

Fallow 1-year (n=3), annuals, ruderals (weedy species)

Fallow 3 year (n=2), herbaceous annuals and perennials, few woody seedlings and saplings

Fallow 7 year (n=2), annuals and perennial herbs, young shrubs and trees (low bush appearance)

Fallow 14 year (n=2), annuals and perennials, young trees and shrubs, several trees with shade canopies.

2.2 Soil sampling and substrate utilization assay

Fifteen cores (each 1.2 cm in diameter x 25 cm deep) were collected from each field site and grouped into five composite samples (3 combined per group) in

April and August 1996. Since there were no significant differences between sampling times (MANOVA analysis) the soils were combined. Cores were taken using a stratified sampling procedure. Each soil core (n=5) was analyzed for microbial and nutrient properties.

To determine if any changes occurred in the physiological abilities (functional diversity) in the microbial community, BIOLOG GN (gram negative) and GP (gram positive) microtiter plates, each containing wells with 96 different C sources and a redox dye (Biolog Inc., Hayward, CA 94545, USA), were used. Combining these two types of plates provide a test with 128 different carbon substrates. Soil samples were mixed with deionized water and the soil solution adjusted to yield cell counts of an initial well density of approximately 30,000. The soil dilution was then micro-pipetted into 96-well microtiter plates, duplicate aliquots were used for each sample. Plates were incubated at 25°C for 24-72 hours, and optically scanned with a plate reader (Biolog, minimum optical density was 405). All calculations presented here are done with readings at 72 hours of incubation.

Microbial biomass was determined using the substrate induced respiration (SIR) method (Anderson and Domsch 1978). A 20 g subsample was used to assess hyphal lengths employing the membrane filter technique using trypan blue as the hyphal stain (Dhillion et al., 1996). Extracellular hydrolytic enzyme assay for dehydrogenase followed Tabatabai (1982), while mycorrhizal spore extraction and staining is described in Dhillion et al. (1996).

2.3 Statistics

For all soil biotic and abiotic properties regressions with age of site as the independent variable were calculated. The activity of the microflora on each of the substrates after incubation may reflect the species composition of the biota and/or the physiological ability of the microbial community to assimilate particular substrates (groups: carbohydrates, carboxylic acids, amines/amides, amino acids, polymers) (Garland and Mills, 1991; Biolog 1993; Ritz et al., 1994; Zak et al., 1994; Insam et al., 1996; Dhillion et al., 1996; Willig et al., 1996). Substrate utilization assays were analyzed using canonical discriminate analysis procedure. Functional diversity of the bacterial communities was characterized by indexes, substrate richness, evenness and diversity (see also Zak et al., 1994). The number of substrates that are used by the bacterial community is a measure of substrate richness (S). Substrate diversity (H') encompasses both substrate richness and substrate evenness and can be quantified by the Shannon's Index (Magurran, 1988) as:

$$H' = - \sum p_i (\ln p_i)$$

where p_i is the ratio of the activity on a particular substrate to the sum of activities on all substrates. Substrate evenness (E) (Magurran, 1988) measures the equitability of the activities across all utilized substrates and is defined as:

$$E = H'/H'_{max} = H'/\ln S$$

3. Results

Organic matter, microbial biomass carbon, the number of mycorrhizal spores, and the density of woody species showed relatively strong positive relationships with the age of fallow (Table 1). Inorganic nutrients N and P showed no relationship while Ca and Mg had weak positive relationships with the age of fallow. The age of fallow sites was positively correlated with the number of reactions and the summation of activities. All diversity indexes show that the younger fallow sites were more alike than the older ones, having lower substrate diversity, richness and evenness (Table 2). There were no significant differences between the cropped sites and the one year old site for all the indexes, whereas soil from the 3 year old fallow sites was significantly different from both the younger and older fallow sites. Although substrate richness was significantly higher in soils of the 7 year fallow than the 14 year fallow, substrate diversity tended to be higher while substrate evenness was significantly higher in the 14 year fallow.

Table 1 Relationships of age of fallow and selected soil properties (data soil cores) in a chronosequence of fallow communities. For details on site vegetation and sample size see section 2. Relationship positive (+) or none (/) between age of fallow and variable.

Variable	Age of Fallow r^2	p	Relationship
Organic Matter (%)	0.73	0.005	+
N (mg kg^{-1})	0.06	0.45	/
P (Bray) (mg kg^{-1})	0.27	0.052	/
Ca (mg kg^{-1})	0.53	0.01	+
Mg (mg kg^{-1})	0.61	0.01	+
Microbial biomass-C (μg CO_2 g^{-1}soil h^{-1})	0.75	0.005	+
Hyphal length (cm g^{-1} soil)	0.32	0.03	+
Dehydrogenase activity (μg g^{-1} soil h^{-1})	0.54	0.01	+
Mycorrhizal spores (no. g^{-1} soil)	0.67	0.02	+
Density of woody species (number 10 m^{-2})	0.74	0.001	+
Density of perennial species (number m^{-2})	0.43	0.04	+
Number of reactions (S) Biolog substrate richness	0.68	0.005	+
Summation of activities Biolog	0.72	0.005	+

The canonical discriminant analysis was able to separate the fallow sites by age (Fig. 1) with two functions. Amino acids like L-leucine, L- phenylalanine, and L-ornithine among other substrates of different groups were positively correlated with function 1 (Table 3). On the other hand the carboxylic acids, citric acid and γ-hydroxybutyric acid were positively and L-rhamnose and glucose-6-phosphate were negatively correlated with function 2.

Table 2 Mean values of microbial functional diversity, classified by land use (age of fallow and agricultural fields). Means in a row with similar letters are not significantly different (p > 0.05). For details site vegetation and sample size see section 2.

Indexes	Land use				
	Crop	Age of fallow (years)			
		1	3	7	14
Substrate Richness (S)	64a	70a	84b	106d	97c
Substrate Diversity(H')	1.48a	1.52a	1.64b	1.83c	1.85c
Substrate Evenness(E)	0.81a	0.83a	0.87b	0.93c	0.96d

Table 3 Correlations between discriminating functions and substrates.

Carbon Source	Function 1	Function 2
L-leucine	0.58	
succinamic acid	0.57	
glucose-1-phosphate	0.53	
L-phenylalanine	0.46	
p-methyl-D-glucoside	0.40	
Inosine	0.40	
L-ornithine	0.38	
citric acid		0.51
γ-hydroxybutyric acid		0.46
L-rhamnose		-0.35
glucose-6-phosphate		-0.31

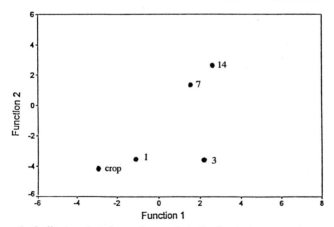

Figure 1 Ordinate plot from the canonical discriminant analysis of the substrate utilization assay. Group means are labeled to show the ages (1, 3, 7, 14) of fallow and cropped sites. The greater the distance of the group means the more different are the substrate use abilities of the bacterial communities of soils from the chronosequence of fallow. No differences were found among samples within sites (variability was tested using nested ANOVA).

4. Discussion

The age of fallow showed strong relationships with several soil properties, nutrient and biotic activities, including an assessment of microbial functional abilities. Previous studies on age of sites and land practice have shown strong relationships of microbial biomass-C and organic matter (e.g., Insam and Domsch, 1988; Insam, 1990) although few studies have simultaneously evaluated other parameters shown here. Microbial biomass carbon indicates changes in the carbon status of soils. The data here supports the view that soil properties, biotic in particular, improve with age of fallow via several possible mechanisms, mediated by the establishment of perennial herbaceous/woody species, and the interactions of inputs, presumably via roots, with soil biota and nutrients (see non-fallow studies, Insam et al., 1996; Dhillion et al., 1996; Willig et al., 1996). All relationships suggest that there may be differences in resources available (e.g., organic matter) along the chronosequence of fallow, reflected in part by the enhanced enzymatic and microbial functional activities. Although plant diversity was not determined, changes in plant functional groups (herbaceous annuals/perennials, wood species) point towards positive influences of these groups in building ecosystem soil properties. Age of abandoned agricultural sites, under slash and burn systems, in Thailand and Malaysia also tend to have increased rates of organic matter turnover, enzymatic activities, microbial and fungal biomass, and root biomass (work in progress). Certainly the successional trajectory followed by abandoned sites depends on prior agricultural history, residual seed bank, and the status of neighboring fields.

Other workers have also hypothesized that microbial functional abilities vary as a consequence of differential resource input and availability as a result of direct or indirect changes in vegetation composition or dynamics caused by climate conditions, disturbance intensities or land use practices (e.g., Dhillion et al. 1996; Insam et al. 1996; Willig et al. 1996; see studies in Ritz et al. 1994 and this volume). In experiments manipulating plant diversity and those under elevated CO_2 levels, root biomass and length were shown to increase significantly in concert with the enhancement of microbial activities (Roy et al. 1995; Dhillion et al. 1996; Roy and Dhillion, unpublished data). Although the Biolog assay does not determine the proportion of the total bacterial community responsible for the carbon utilization patterns it does provide an assessment of its impact on a suite of resources. Certain groups of substrates may be significant to the understanding of the types of changes occurring in the soil, for example the amino acids which separate well among sites (Zak et al. 1994; Dhillion et al. 1996). Overall the practice of fallow ought to be encouraged but precise management recommendations will require more detailed and long term studies. This study shows that fallow increases plant functional diversity, microbial activities, and nutrient levels, which indirectly contribute towards their economic value.

Studies using the BIOLOG technique have shown differences in bacterial utilization of substrates among soils of grass and shrub communities, communities under elevated CO_2 levels, in the rhizosphere of different tree species, under grazing regimes, natural and anthropogenic disturbances, or during decomposition (e.g., Garland and Mills 1991; Grayston *et al.*, 1994; Zak *et al.*, 1994; Dhillion *et al.* 1996; Willig *et al.* 1996). Given the methodological and time constraints for evaluating genetic and taxonomic diversity among microbes, assessment of physiological abilities or functional diversity of microbial communities provides a viable source of information pointing to linkages and roles of microbe activity to other species diversity, land-use or climatic changes (e.g., Bochner 1989; Hooper *et al.* 1995; Roy *et al.* 1995; Zak *et al.* 1994). Although the differences observed in the above studies and this one may be most likely attributed to organic matter input/changes, via plant diversity changes, litter quality and exudates, this study reports for the first time influences of a traditional land use practice on microbial functional diversity. It must be noted that even when microbial biomass, turnover and overall activity parameters do not show shifts or differences among treatments the functional abilities of the microbial communities may well do so (Insam *et al.*, 1996; Dhillion *et al.* 1996). The importance and relevance of conducting sophisticated analyses for more detailed microbial identities are not undermined by the current method. This method, by providing rapid initial insight to the functional abilities of the microbes, may be seen as a tool to point towards the potential areas of exploration where costly methods can be utilized. Furthermore this study provides evidence that the simultaneous study of soil nutrients, vegetation composition and inputs, and soil microbial activity is essential in understanding processes associated with land use practices like fallow.

Acknowledgments. This research was in part supported by ORSTOM, ICRISAT, and the UNESCO-MAB program on Fallow. Logistic support from the Department of Biology and Nature Conservation, The Agricultural University of Norway is acknowledged.

5. References

Anderson JPE, Domsch KH (1978) A physiological method for the quantificative measurement of microbial biomass in soils. Soil Biol Biochem 10:215-221

Atlas R (1984) Diversity of microbial communities. Adv Microb Ecol 7: 1-47

Biolog (1993) Instructions for use of the Biolog GP and GN Microplates. Biolog Inc, Hayward, Calif

Bochner B (1989) Instructions for the use of the Biolog GP and GN Microplates. Biolog Inc, Calif

Dhillion S, Roy J, Abrams M (in press) Assessing the impact of elevated CO_2 on soil microbial activity in a Mediterranean model ecosystem. Plant and Soil

Floret C, Pontanier R (eds) (1993) La jachere an Afrique tropicale. ORSTOM, Paris

Garland JL, Mills AL (1991) Classification and characterization of heterotrophic microbial communities on the basis of patterns of community-level sole-carbon-source utilization. Appl Environ Microbiol 57: 2351-2359

Grayston SJ, Campbell CD, Vaughan D (1994) Microbial diversity in the rhizospheres of different tree species. In: Parkhurst CE, Doube BM, Gupta WSR, Grace (eds) Soil Biota - Management in Sustainable Farming Systems, CSIRO Press, Adelaide: 155-157

Heywood VH, Watson RT (eds) (1995) Global Biodiversity Assessment. Cambridge Univ Press

Hooper D, Hawksworth D, Dhillion S (1995) Microbial diversity and Ecosystem Processes. In: Global Biodiversity Assessment. Heywood VH, Watson RT, (eds). Cambridge Univ. Press

Insam H, Domsch KH (1988) Relationship between soil organic carbon and microbial biomass on chronosequences of reclamation sites. Microbial Ecology 15:177-188

Insam H (1990) Are the soil microbial biomass and basal respiration governed by the climatic regime. Soil Biol Biochem 22:525-532

Insam H, Rangger A, Henrich M, Hitzl W (1996) The effect of grazing on soil microbial biomass and community on alpine pastures. Phyton, 36: (205-216).

Magurran AE (1988) Ecological diversity and its measurement. Princeton University Press, Princeton, New Jersey

Mooney HA, Lubchenco J, Dirzo R, Sala OE (1995) Biodiversity and ecosystem functioning: ecosystem analyses. In: Heywood VH, Watson RT (eds) Global Biodiversity Assessment. Cambridge Univ Press

Parkinson D, Coleman DC (1991) Microbial communities, activity, and biomass. Agric Ecosyst Environ 34:3-33

Ricklefs RE, Schluter DS (1993) Species Diversity in Ecological Communities: Historical and Geographical Perspectives. The University of Chicago Press, Chicago

Ritz K, Dighton J, Giller KE (eds) (1994) Beyond the biomass. Wiley-Sayce Publication

Roy J, Guillerm J-L, Navas M-L, Dhillion S (1995) Responses to elevated CO_2 in Mediterranean old-field microcosms: species, community and ecosystem components. In Carbon dioxide, populations and communities. Körner Ch, Bazzaz FA (eds). Physiological Ecology Series, Academic Press, San Diego

Tabatabai MA (1982) Soil enzymes. In Methods of soil Analysis. Part 2. Page AL (ed) Agron Monogr, ASA and SSSA, Madison, Wisconsin

Willig MR, Moorhead DL, Cox SB, Zak J (in press) Functional diversity of soil bacterial communities in the Tabonuco forest: the interaction of anthropogenic and natural disturbance. Biotropica

Zak JC, Willig M, Moorhead D, Wildman H (1994) Functional diversity of microbial communities: a quantitative approach. Soil Biol Biochem 26:1101-1108

Substrate Utilization Patterns of Extractable and Non-Extractable Bacterial Fractions in Neutral and Acidic Beech Forest Soils

Silke Kreitz and Traute-Heidi Anderson

Institut für Bodenbiologie, Bundesforschungsanstalt für Landwirtschaft, Bundesallee 50, 38116 Braunschweig, Germany

Abstract. Substrate utilization patterns of four neutral and four acidic beech forest soils were examined using BIOLOG GN microplates. Soil bacteria were extracted using a three batch fractionated centrifugation procedure that produces two fractions. One contains the extracted and one the non-extractable bacteria and the fungi. Substrate utilization patterns of both fractions were determined with inhibition of fungal growth. All samples cluster distinctly into four different groups which correspond to the two fractions of the acidic and the neutral soils. Bacteria in acidic soils have generally a lower overall colour development indicating lower utilization activity. The extracted bacteria from acidic soils show a lower functional diversity than those extracted from neutral soils, but no difference in diversity was observed when the results of the extracted and the non-extractable bacteria of each soil were combined. In acidic soils substrate utilization abilities are heterogenously distributed between the two fractions. This effect mainly occurs in carbohydrate utilization. Bacterial communities in acidic soils are especially unable to degrade carboxylic acids. As decarboxylation is a main process of the initial decomposition of organic residues, this inability may be one reason for the retarded degradation of freshly fallen litter in acidic soils.

Key words. BIOLOG GN microplates, beech forest soils, acidification, functional diversity, decarboxylation.

1. Introduction

In forest soils, acidity affects several ecological processes including the solubility and exchange reactions of plant nutrients and toxic metals, soil biological activity and soil mineral weathering (Binkley and Richter, 1987). Soil acidity is generally linked with an increased amount of biologically toxic cations such as Al^{3+} and Mn^{2+} (Alexander, 1980; Foy, 1984), reduced microbial biomass (Alexander, 1980; Wolters and Joergensen, 1991) and decreased rate of organic matter decomposition (Alexander, 1980, Zelles et al., 1987). The decreased decomposition rate is especially evident for freshly fallen litter (Wolters and Joergensen, 1991; Motavalli et al., 1995) and thus may be partly responsible for the accumulation of organic compounds on the soil surface and C depletion in the mineral soil of acidic forest soils (Wolters and Joergensen, 1991).

In our approach, we examined substrate utilization and functional diversity of acidic and neutral beech forest soils that developed under identical environments (climate, topography, vegetation) but from different parent materials. Under these conditions the performance of soil organisms can hypothetically be related to the process of soil development (i.e. nutrient loss and acidification) (Jenny, 1941; Wolters and Joergensen, 1991). Functional diversity of a microbial community describes the potential of this community to degrade different organic compounds (Zak et al., 1994). It is interrelated, but not comparable to genetic or taxonomic diversity (Solbrig, 1991). Using BIOLOG GN microplates (Garland and Mills, 1991), we intended to show if substrate utilization patterns of acidic and neutral beech forest soils could be classified into different groups and to examine pH-related differences in overall activity, functional diversity and fractionation (i.e. heterogeneity in utilization abilities of extractable and non-extractable soil bacteria).

2. Materials and Methods

2.1 Soils

All soils were collected in beech forests around Braunschweig (Lower Saxony, Germany). Soil types and characteristics are shown in Table 1. Soil samples (Ah-horizons, 0-5 cm) were taken exclusively in early spring (beginning of April) before the emergence of ground-cover. Soils were stored unsieved at 4°C to minimize microbial death during storage. Before experiments, the soils were sieved (2 mm) and adjusted to a moisture of 50-60% of their maximal water holding capacity.

2.2 Extraction of bacteria from soil

Soil bacteria were extracted from soil using a modified three batch procedure (Hopkins et al., 1991). 5 g fresh weight soil were blended (Waring blender) with 50 ml 0.1% sodium cholate /2.5% polyethylene glycol 6000 (PEG) solution for 1 min while cooling with water. 15 g cation exchange resin (Dowex 50WX8, 50-100 mesh, Sigma) were added and the suspension was shaken at 4°C for 90 min. Afterwards, the resin was eliminated by wet sieving (mesh size 105 μm) and released bacteria separated from undispersed soil particles and fungi by low speed centrifugation (10 min, 750 g). The supernatant was collected, the pellet was resuspended in 40 ml 0.1% sodium cholate/2.5% PEG solution and ultrasonicated for 4 min in an ultrasound bath.

After another low speed centrifugation the supernatant was collected and the pellet was ultrasonicated in 30 ml 1 mM ethylene-diamide-tetraacetic-acid (EDTA) and afterwards shaken (15 min, 4°C). The pellet of the subsequent low speed centrifugation was resuspended in 20 ml sterile physiological NaCl solution. This fraction is subsequently called "pellet" and contains the non-extractable bacteria, the fungi and coarser soil particles. The supernatants of the three low speed centrifugations were combined and concentrated by high speed centrifugation (30 min, 10 000 g). The supernatant was discarded and the pellet resuspended in 5 ml sterile physiological NaCl solution. This fraction is subsequently called "extract" and contains the extracted bacteria and fine dispersed soil particles.

Table 1. Soil characteristics

Soil		Soil type[1]	pH [KCl]	C_{mic} [mg/g]	C_{org} [%]	C/N-ratio
acidic soils	aI	Fen over Chalkmarl	2.9	0.30	11.5	24.3
	aII	podsolic Brown Earth	3.0	0.49	12.9	23.8
	aIII	podsolic Brown Earth	3.1	0.26	5.4	23.8
	aIV	Parabrown Earth	3.5	0.41	6.8	21.1
neutral soils	nI	alkaline Brown Earth	7.2	0.66	2.4	13.3
	nII	Rendzina-Brown Earth	6.9	1.46	6.4	14.4
	nIII	podsolic Parabrown Earth	6.7	1.66	7.3	21.2
	nIV	Brown Earth-Pseudogley	7.3	1.03	4.4	n.d.[2]

[1] German classification system (AK Bodensystematik, 1985)
[2] not determined

2.3 Sample preparation, inoculation and incubation of BIOLOG microplates

From each soil two replicate extraction procedures were performed. Both fractions were diluted 1:200 with sterile Ringer solution of one quarter strength containing 0.1% cycloheximide in order to inhibit fungal growth. This dilution level corresponds to 5 mg ml^{-1} fresh weight soil in extracts and 1.25 mg ml^{-1} fresh weight soil in pellets. Background absorbances of both extracts and pellets of one soil were similar. Each diluted fraction was inoculated into a BIOLOG GN microplate (BIOLOG Inc., Hayworth, CA) (150 μl/well) and incubated at 22°C. Colour development was measured photometrically immediately after inoculation and after 16, 20, 24, 28, 40, 44, 48 and 65 h. Although the peak absorbance of the tetrazolium dye occurs at 590 nm we used absorbance at 550 nm because our plate reader was equipped with only that filter. In that our measures of activity are consequently conservative, we can be confident of site differences detected by our technique.

2.4 Data analysis

Absorbance values for the wells were first blanked against the background absorbances measured directly after inoculation. Secondly the absorbance of the control well was subtracted from that of each well containing a C-source. All absorbance values below 0.05 were considered as 0 in subsequent data analyses. Average well colour development (AWCD) was calculated from each plate at each time according to Garland and Mills (1991).

Because the number of utilized substrates and the AWCD strongly depend on incubation time, for each plate a point of time had to be chosen, when both measures are characteristic for the inoculated bacteria and comparable to the results of other plates. We decided to use the absorbance pattern of each microplate after that incubation time, when the number of utilized substrates had reached a saturation and the AWCD was still increasing (Fig. 1). At this point of time further incubation would not result in significantly higher numbers of utilized substrates and the absorbances of each substrate are mostly proportional to their degree of utilization.

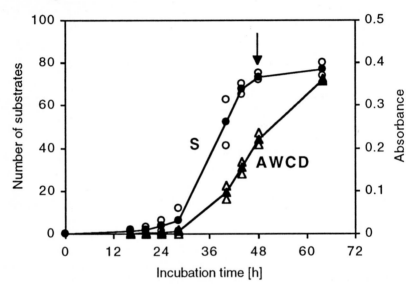

Fig. 1. Dependency of the number of utilized substrates (S) (●) and AWCD (▲) from incubation time. Extracts of soil nIII. The arrow indicates the optimal incubation time, when S reaches a saturation but AWCD is still increasing (see text). Values of two replicate experiments (open symbols) and the mean of both replicates (closed symbols).

The absorbance values of each plate at its optimal incubation time were used to perform a principal component analysis (PCA) in order to classify substrate utilization patterns (Garland and Mills, 1991; Zak et al., 1994) and to calculate the Shannon-Weaver index to get a measure of functional diversity (Zak et al., 1994).

To evaluate substrate utilization of the whole soil bacterial community (i.e. extractable and non-extractable bacteria), the absorbance values of the two plates inoculated with the extract and pellet of each extraction procedure were added and AWCD, number of utilized substrates and Shannon index were calculated on the basis of this new data set.

Differences between classified groups were evaluated by calculating the means and standard deviations for AWCD, number of utilized substrates and Shannon index of each group. Significant differences between groups were tested using the two tailed t-test.

3. Results

The similarity of utilization patterns as provided by soil fractions of replicate extractions at their optimal incubation time was in the range between 93 and 99 % using the Morisita-Horn index (Wolda, 1981), which is a quantitative similarity index and thus based on quantitative absorbance patterns rather than binary substrate utilization.

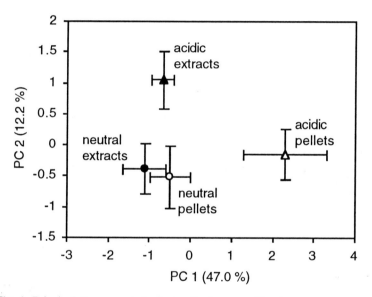

Fig. 2. Principal Component Analysis of substrate utilization patterns of all samples at their optimal incubation time. Means and standard deviation (n=8) or the four examined groups are shown.

By PCA distinctive substrate utilization patterns were found for acidic extracts, neutral extracts, acidic pellets and neutral pellets (Fig. 2). The patterns of the extracts and pellets of the neutral soils were quite similar. Their variability is mainly described by the first principal component (PC). Conversely, the utilization patterns of the acidic extracts and pellets vary strongly along both, first and second, PCs. The extracts of the acid and neutral soils also produce distinctive, non overlapping utilization patterns, which are divided mainly along the second PC axis (Fig 2).

The comparison of the mean AWCD, number of utilized substrates and Shannon index of acidic and neutral extracts shows that bacteria in acid extracts have a lower overall activity, utilize fewer substrates and are less diverse (Fig. 3). The differences are slight, but significant. Regarding the three main substrate groups provided by the BIOLOG GN microplates (carbohydrates, carboxylic acids and amino acids), these effects are especially evident for the carboxylic acids, which differ highly significantly between both groups. The other substrate groups also contribute to these effects, but to a lesser extent. They differ at a significance level of $p < 0.05$. No statistical differences were observed in the utilization activity of the amino acids and the number of utilized carbohydrates (Fig. 3).

Referring to the combined utilization patterns of both extractable and non-extractable bacteria, the comparison of the mean AWCD, number of utilized substrates and Shannon index of acidic and neutral soils lead to a different result (Fig. 4). Only the overall utilization activity is much lower in acidic soils, but no difference was observed concerning the number of utilized substrates and functional diversity. On the other hand, the reduced ability of bacterial communities in the acidic soils to utilize carboxylic acids is still evident (Fig. 4).

As shown before, using PCA (Fig. 2), the substrate utilization patterns of bacterial communities in the acidic extracts and pellets are much more dissimilar than those of the neutral extracts and pellets. Regarding the bacteria in neutral extracts and pellets, communities in both fractions utilize mainly the same substrates (Fig. 5). Conversely, the bacteria in the acidic pellets are able to degrate some substrates that are not utilized by the bacteria in the extracts as indicated by arrows in Fig. 6. This is the case for 5 carbohydrates, 1 carboxylic acid and 2 amino acids (Fig. 6, Table 2).

Table 2. Fractionation according to substrate utilization in acidic soils. Substrates only utilized by bacterial communities in pellets. These substrates are marked in Fig. 6. with an arrow.

Substrate group	Substrate	Substrate group	Substrate
Carbohydrates	lactulose	Carboxylic acids	D-glucosaminic acid
	α-D-lactose gentiobiose	Amino acids	L-threonine
			L-phenylalanine
	β-methyl-D-gluco-side		
	D-melibiose		

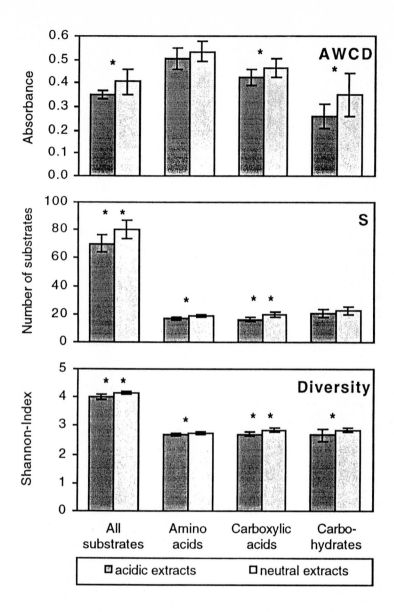

Fig. 3. Overall activity (AWCD), number of utilized substrates (S) and functional diversity of acidic and neutral extracts. Means and standard deviation of each group (n=8) are plotted for all substrates and separately for amino acids, carboxylic acids and carbohydrates. * significant difference between groups with p < 0.05; ** significant difference between groups with p < 0.01.

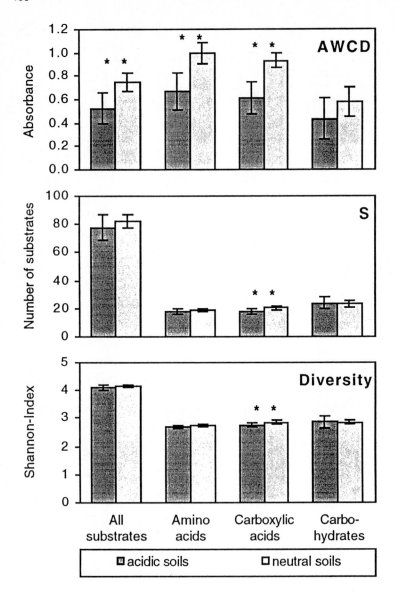

Fig. 4. Overall activity (AWCD), number of utilized substrates (S) and functional diversity of the combined utilization patterns of both fractions of acidic and neutral soils. Means and standard deviation of each group (n=8) are plotted for all substrates and separately for amino acids, carboxylic acids and carbohydrates. * significant difference between groups with $p < 0.05$; ** significant difference between groups with $p < 0.01$.

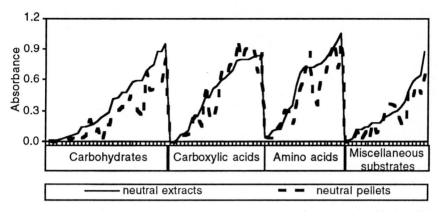

Fig. 5. Substrate utilization of neutral extracts and pellets. The mean absorbances of each group are plotted for each substrate. Substrates are grouped into carbohydrates, carboxylic acids, amino acids and miscellaneous substrates. Within each substrate group, substrates are ranked according to increasing absorbances of the extract fractions.

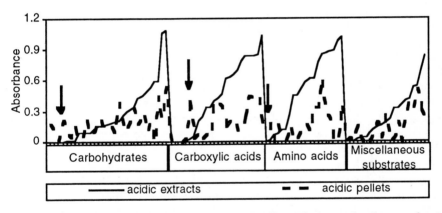

Fig. 6. Substrate utilization of acidic extracts and pellets. The mean absorbances of each group are plotted for each substrate. Substrates are grouped into carbohydrates, carboxylic acids, amino acids and miscellaneous substrates. Within each substrate group, substrates are ranked according to increasing absorbances of the extract fractions. The arrows indicate those substrates that are only utilized by bacteria in pellets (see text and Table 2).

4. Discussion

The results of this study indicate that bacterial communities of acidic and neutral beech forest soils produce distinctive patterns of C-source utilization based on the optimal incubation time of each soil. We found this reference more suitable for our approach than an equivalent AWCD. It is uncertain if these different utilization patterns are evoked by differences in community structure or changes in decomposition abilities caused by acidic soil conditions. Both changes in enzyme activity (Dick *et al.*, 1988; Haynes and Swift, 1988) and the composition of the microbial population (Shah *et al.*, 1990; Nodar *et al.*, 1992) were observed according to soil acidification. However, to understand mechanisms in decomposition processes of organic matter it is not necessary to know the bacterial community structure on its taxonomic level but rather its functional potential.

The reduced AWCD of BIOLOG plates inoculated with acidic soil fractions, which is a measure for a reduced overall utilization activity of bacterial communities in acidic soils, may be caused by the smaller inoculum size of the acidic soil fractions. As the dilutions of the fresh weight soil of each fraction are equal, the actual number of inoculated bacteria varies according to the amount of soil microbial biomass. We decided to compare substrate utilization pattern on the basis of the amount of fresh weight soil in inoculated suspension rather than equal numbers of inoculated bacteria (Garland and Mills, 1991) or equal activity indicated by an AWCD threshold (Garland, 1996). Both, reduced biomass (Alexander, 1980; Wolters and Joergensen, 1991) and reduced decomposition activity (Alexander, 1980; Jenkinson, 1981, Zelles *et al.*, 1987) are characteristic for bacterial communities in acidic soils. Thus, these parameters should not be eliminated when differential substrate utilization in acidic soils is examined.

The reduced AWCD of the acidic soil fractions may also be caused by an increased pH in the BIOLOG microplates compared to the initial pH of the soils. Since in the extraction procedures as well as during the dilution steps all used solutions are unbuffered, pH-differences between acidic and neutral soils are maintained during incubation in the BIOLOG microplates. Nevertheless, the use of cation exchange resins in the extraction procedure causes for all soils an increase of the pH for about one unit. It is unlikely that differences in overall acitivity are caused by these pH changes because pH tolerances of bacterial communities are usually not that narrow. This is especially the case for bacteria in acidic soils because due to the logarithmic nature of the pH the actual decrease in H^+-concentration with increasing pH is less at acidic than at neutral pH. Additionally, at the low pH 3, which is found for the acidic soils, a pH-increase would cause an increase in activity rather than a decrease. For that reason, differences in overall acitivity between acidic and neutral soils are unlikely to be overestimated.

Bacteria in acidic soils have different utilization activities and abilities in soil fractions, the non-extractable fraction of the whole bacterial community has a lower overall activity and another degradation potential than the extractable bacteria. As clay particles have negative surface charges, H^+ of acid soils mainly accumulate at clay surfaces. Thus, the actual pH outside the soil aggregates is much lower than in the surrounding solution (Alexander, 1980). Soil bacteria are mainly attached to clay surfaces or live within aggregates, just a minor portion is floating freely in the soil solution (Hattori and Hattori, 1976). Therefore, bacteria within aggregates are less exposed to harmful environmental conditions than outside. It is not proven if these bacteria belong to different taxa, but at least they have different functional potentials. Microorganisms in soils are heterogeneously clustered into different microhabitats (Hattori and Hattori, 1976). The more these microhabitats are unequal according to their physical and chemical conditions, the more do the taxonomic and functional characteristics of the inhabiting microorganisms vary. In case of the investigated acidic soils the functional variation between bacteria living within and outside of soil aggregates, i.e. extractable and non-extractable bacteria, consists in their ability to utilize some carbohydrates, carboxylic and amino acids only by the non-extractable bacteria. The most of these substrates (5) are carbohydrates.

Bacteria in both acidic fractions show a lack in degradation of carboxylic acids. As decarboxylation of organic acids is discussed as one of the major steps in the initial decomposition of freshly fallen litter (Yan *et al.* 1996), this effect may contribute to the known accumulation of organic matter on soil surfaces (Wolters and Joergensen, 1991).

Concerning the functional diversity, we found that in acidic soils it is lower if only the extractable bacteria are considered but it does not differ between acidic and neutral soils for bacteria of both soil fractions taken together. Thus, if one aims to investigate bacterial communities of different soils, it is of utmost importance to consider the fraction of the whole bacterial communities that are inoculated into the BIOLOG plates and to take the heterogenous distribution of bacteria into account when interpreting the results obtained using the BIOLOG-procedure. The usefulness of a method depends not only on the possibilities and limitations of the method itself but more so on the questions asked. For example, one can argue that decomposition of organic matter in soils is mainly caused by those bacteria which are in early contact with the degradable organic compounds. These bacteria live outside the aggregates and thus mainly belong to the extractable fraction. Therefore, to understand the decomposition properties of a soil, only the knowledge of the functional diversity and abilities of the extractable bacteria is important. On the other hand, if one is interested in the whole degradation potential of a soil, it is necessary to consider a far bigger proportion of the soil bacterial community, which is in the case of this study provided by including the utilization patterns of the non-extractable bacteria.

5. References

AK Bodensystematik (1985) Soil classification of the Federal Republic of Germany (abridged English version). Mitteilungen der Deutschen Bodenkundlichen Gesellschaft 44: 1-96
Alexander M (1980) Effects of acidity on microorganisms and microbial processes in soils. In: Hutchinson TC, Havas M (eds) Effects of acid precipitation on terrestrial ecosystems. Plenum Press, New York, pp 363-374
Binkley D, Richter D (1987) Nutrient cylces and H^+ budgets of forest ecosystems. In: MacFadyen A, Ford ED (eds) Advances in ecological research, Vol. 16. Academic Press, Orlando, pp 1-51

Dick RP, Rasmussen PE, Kerle EA (1988) Influence of long-term residue management on soil enzyme activities in relation to soil chemical properties of a wheat-fallow system. Biol Fertil Soils 6: 159-164

Foy CD (1984) Physiological effects of hydrogen, aluminium and manganese toxicities in acid soil. In: Adams F (ed) Soil acidity and liming. Am Soc Agron, Madison, pp 57-97

Garland JL, Mills AL (1991) Classification and characterization of heterotrophic microbial communities on the basis of patterns of community-level sole-carbon-source utilization. Appl Environ Microbiol 33: 434-444

Garland JL (1996) Analytical approaches to the characterization of samples of microbial communities using patterns of potential C source utilization. Soil Biol Biochem 28: 213-221

Hattori T, Hattori R (1976) The physical environment in soil microbiology: an attempt to extend principles of microbiology to soil microorganisms. Crit Rev Microbiol 4: 423-461

Haynes RJ, Swift RS (1988) Effects of lime and phosphate additions on changes in enzyme acitvities, microbial biomass and levels of extractable nitrogen, sulphur and phosphorus in an acid soil. Biol Fertil Soils 6: 153-158

Hopkins DW, Macnaughten SJ, O'Donnell AG (1991) A dispersion and differential centrifugation technique for representatively sampling microorganisms from soil. Soil Biol Biochem 23: 217-225

Jenny (1941) Factors of soil formation. McGraw-Hill, New York

Motavalli PP, Palm CA, Parton WJ, Elliott ET, Frey SD (1995) Soil pH and organic C dynamics in tropical forest soils: evidence from laboratory and simulation studies. Soil Biol Biochem 27:1589-1599

Nodar R, Acea MJ, Carballas T (1992) Microbiological response to CA(OH)$_2$ treatments in a forest soil. FEMS Microbiol Ecol 86: 213-219

Shah Z, Adams WA, Haven CDV (1990) Composition and activity of the microbial population in an acidic upland soil and effects of liming. Soil Biol Biochem 22: 257-263

Solbrig OT (1991) From genes to ecosystems: a research agenda for biodiversity. Report of a IUBS-SCOPE-UNESCO workshop. The International Union of Biological Sciences, 51 Boulevard de Montmorency, Paris, France

Wolda H (1981) Similarity indices, sample size and diversity. Oecologia 50: 296-302

Wolters V, Joergensen RG (1991) Microbial carbon turnover in beech forest soils at different stages of acidification. Soil Biol Biochem 23: 897-902

Yan F, Schubert S, Mengel K (1996) Soil pH increase due to biological decarboxylation of organic anions. Soil Biol Biochem 28: 617-624

Zak JC, Willig MR, Moorhead DL, Wildman HG (1994) Functional diversity of microbial communities: a quantitative approach. Soil Biol Biochem 26: 1101-1108

Zelles L, Scheunert I, Kreutzer K (1987) Effect of artificial irrigation, acid precipitation and liming on the microbial activity in soil of a srpuce forest. Biol Fertil Soils 4: 137-143

Biolog Metabolic Fingerprints for Clustering Marine Oligotrophic Bacteria from Polar Regions

Tjhing Lok Tan

Alfred-Wegener-Institut für Polar- und Meeresforschung, Am Handelshafen 12, 27570 Bremerhaven, Germany

Abstract. Oligotrophic bacteria from the Western Greenland Sea (Expedition ARKTIS IV/2, 1987) have been isolated by enrichment culture techniques in dialysis chambers or by continuous-flow of seawater through nylon or glass-fibre filter in double Petri dishes. The isolated strains appeared to be Gram-negative and psychrotrophic. Organic substrate utilizations of 52 strains were determined in Biolog microplates for Gram-negatives and the obtained metabolic fingerprints used for clustering the bacteria. Three distinct clusters can be recognized at the 80% similarity level, as recommended by the Biolog company.

These bacteria were compared with 55 oligotrophic isolates from the Gunnerus and Astrid Ridge, Antarctic Ocean (Expedition ANTARKTIS VIII/6, 1990). From the five clusters found among the Antarctic bacteria, two were the same as two clusters from the Arctic. These identical groups of bacteria from the north and south polar regions will be the subject of phylogenic investigations to study the strategies of cold adaptations in bacteria.

Keywords. Arctic, antarctic, biolog, cluster analysis, marine, metabolic fingerprints, oligotrophs

1. Introduction

Investigations of biomasses, activities and community structures of copiotrophic and oligotrophic bacteria were carried out on two expeditions with the RV "Polarstern" in the Arctic (Krause *et al.,* 1989) and Antarctic (Fütterer and Schrems, 1991), regions. Results of bacterial biomasses and nutritional requirements of copiotrophs from Fram Strait and Western Greenland Sea have been published by Tan and Rüger (1991). On the same Arctic expedition, enrichment cultures were performed in dialysis chambers and on membrane filters in so-called "double Petri dishes" in order to isolate cold adapted and oligotrophic bacteria. The results of these enrichment culture techniques for isolating oligotrophic bacteria and the use of the Biolog substrate metabolism system for clustering the marine oligotrophic strains from north and south polar regions are presented here.

2. Materials and Methods

2.1 Sampling methods and locations of bacteriological stations

Sampling methods and locations of bacteriological stations in Fram Strait and Western Greenland Sea were reported in Tan and Rüger (1991).

2.2 Enrichment culture techniques

Microorganisms from 500 ml seawater were filtered on glass fibre prefilter and 0.1 μm nylon membrane filter (Ultipor N_{66} Posidyne, PALL). After filtration, each of the prefilter and nylon filter were put separately on a quartz microfibre filter (STORA Filter Products, Grycksbo, Sweden), laid on the surface of a porous glass-base of a filter crucible (pore width 40-100 μm; diameter 9 cm), which was inside a glass Petri dish (diameter 14 cm) with a cover. In the Petri dish cover were mounted 2 glass tubes, each 4 cm long with an inner diameter of 3 mm. The sterile seawater medium from the reservoir was pumped through the glass tube into the Petri dish. Flow rate was 6.5 ml h^{-1}, and dilution rate of the continuous-flow culture system was 0.083 h^{-1}. The other tube can be used for inflow of gas (e.g. CO_2). On the base of the Petri dish was mounted another tube (inner diameter 3 mm), which was connected to a peristaltic pump, transporting the seawater from the Petri dish to the effluent vessel. A schematic representation of the continuous-flow culture technique is given in Fig. 1. Continuous-flow cultures were begun on board only from 25 m-depth water samples of station numbers 223, 227, 235, 242 and 253. The flow cultures were ended after 4 months, and the filters suspended in 10 ml of oligotrophic seawater medium according to Ishida et al. (1986) in screw-cap Erlenmeyer flasks and the flasks incubated at 2°C on a rotary shaker at about 100 rpm. To kill cyliates and flagellates grazing on bacteria, 0.1 ml of a 1% actidione (cycloheximide: Serva Feinbiochemika, Heidelberg) solution was added. After 3 months of incubation, dilutions were made from cultures showing turbidity by inoculating fresh oligotrophic media each with 2 drops from the first culture. From these dilution cultures, second dilutions were made after another 3 months. Finally, the second dilution cultures were diluted in microtiter-plates, using an eight-channel Electrapette by mixing 20 μl of bacteria suspension with 200 μl of oligotrophic medium per well, as described by Tan et al. (1996). Turbidities were measured after 3 months of incubation at 2°C with an eight-channel photometer at 405 nm (SLT Labinstruments, Crailsheim, Germany). From microtiter-plate wells showing turbidities, a loopful was streaked out on Basal Seawater Agar plates according to Carlucci et al. (1986). Further purification steps, growth temperatures and maintenance of pure strains will be reported by Tan and Rüger (submitted).

From station numbers 223 (25, 200, 1000 m water depths), 227 (25, 200 m), 235 (25, 200 m) and 242 (25, 200, 1000 m) enrichment cultures in dialysis chambers were also carried out on board. A modified version of the dialysis chamber, used for investigating the toxicity of heavy metals on bacteria, as described by Tan (1986), was applied for enrichment of oligotrophs. Further processing steps were quite similar to the continuous-flow cultures presented above.

2.3 Gram-type of cell wall determinations

The Gram-type of the isolates was determined with the Bactident L-alanine aminopeptidase test strips (E. Merck, Darmstadt, Germany), according to the procedure recommended by the manufacturer.

2.4 Substrate utilizations in Biolog GN MicroPlates

Substrate utilizations were determined in Biolog GN MicroPlates as described by Rüger and Krambeck (1994). However, the cells were suspended in $NaCl-MgCl_2$ solution, made up with 1 n NaOH to a pH of 7.8. The pH decreased to about 7.0 after sterilization in the autoclave. Sodium hydroxide addition is necessary, since the $NaCl-MgCl_2$ solution has a low pH of 5.82. Formazan formations were measured at 590 nm with an eight-channel photometer (SLT Labinstruments) after 1, 2 and 3 d of incubation at 20° or 12°C. Cluster analysis of the metabolic profiles was performed with a Fortran Program-System, modified for application on personal computers by Krambeck and Witzel (1983). The complete linkage algorithm was chosen for clustering, because the presentation of substrate utilization data are more structured, showing more groups and more subgroups, as treated by Sneath and Sokal (1973).

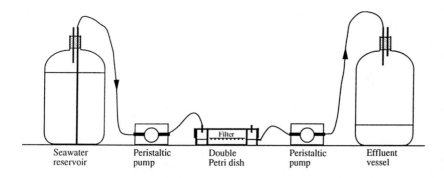

Seawater reservoir   Peristaltic pump   Double Petri dish   Peristaltic pump   Effluent vessel

Fig. 1. Schematic representation of the continous-flow culture technique

3. Results

In total, 112 Arctic strains of marine oligotrophic bacteria could be isolated and kept in pure culture on Basal Seawater Agar and Peptone-Yeast extract-Glucose-Agar slants, as described by Tan and Rüger (submitted). The origin of the 52 strains used in Biolog substrate metabolism studies are compiled in Table 1. Most of the isolates, 38 strains, originated from enrichment cultures in dialysis chambers, and the others were from continuous-flow cultures in "double Petri dishes". But only from glass fibre filters, kept in "double Petri dishes", could bacteria be isolated. There was hardly any growth on the nylon filters, and no turbidities were found in dilution cultures in microtiter-plates. All isolates were Gram-negative and psychrotrophic, having optimum growth temperatures of between 20 and 30°C, but growth occurred even at 37°C or higher.

Table 1. Origin of bacterial strains from the Western Greenland Sea

Station No.	Water depth [m]	Position of Station	Enrichment Technique	Strain No. ARK
223	25	75°33.3'N; 08°48.8'W	Dialysis	176; 179; 180; 184; 186; 188; 189; 190; 193; 194; 195; 196; 197; 199; 201; 204; 205; 206; 212; 213; 215; 217; 219; 222
223	200		Dialysis	164; 165; 166; 167; 168; 169; 170; 171
223	1000		Dialysis	172; 173
235	25	75°09.4'N; 12°27.6'W	DoublePetri	105; 107; 109; 111; 112; 113; 115; 116; 119; 120; 122; 124
235	25		Dialysis	133; 137; 138
242	25	71°56.1'N; 08°21.1'W	DoublePetri	129; 131
242	200		Dialysis	145

Substrate utilization (95 different substrates and a control blank) of 52 Arctic and 55 Antarctic oligotrophs was determined in Biolog GN MicroPlates and the metabolic profiles were used for clustering the bacterial strains. As shown in the dendrogram, six distinct clusters can be recognized at the 80% similarity level (Fig. 2). Substrate metabolic fingerprints of Antarctic strains belonging to cluster IV, V and VI will be reported in Tan and Rüger (submitted). Organic substrates metabolized by the oligotrophs belonging to cluster I (22 Arctic and 7 Antarctic strains), cluster II (21 Arctic strains) and cluster III (9 Arctic and 15 Antarctic strains) are presented (Table 2). In the footnote are listed 23 substrates not utilized by the 74 isolates of cluster I, II and III.

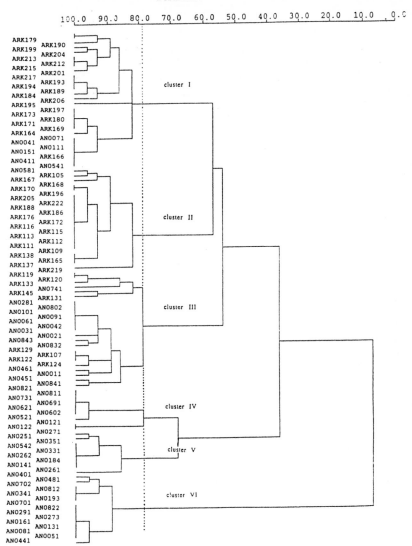

% SIMILARITY

Fig. 2. Dendrogram of arctic and antartic bacterial strains

Colonies on agar media from bacteria belonging to cluster I were of milky-white appearance, moist glistening, convex; cells were rod-shaped, measuring 0.9 µm by 1.35 to 4.0 µm. Bacteria from cluster II formed milky-white, convex, moist glistening colonies on agar media; cells were short to long rods, occurring mostly singly or occasionally in short chains; rods measured 0.9 µm by 1.35 to 4.5 µm.

Bacteria from cluster III showed yellow to orange colonies on agar media; in younger cultures, cells appeared to be short to long rods or filamentous cells,

166

measuring 0.9 μm by 1.8 to 9.0 μm, but in older cultures mostly short to long rods occurred.

Table 2. Substrates metabolized by Arctic-Antarctic strains from cluster I, II and III

	Positive: +	Variable: v	Negative: -

SUBSTRATE	CLUSTER I (29 strains)	CLUSTER II (21 strains)	CLUSTER III (24 strains)
dextrin	+	-	-
glycogen	+	-	-
tween 40	+	+	+
tween 80	+	+	+
N-acetyl-D-glucosamine	-	-	+
adonitol	v	-	-
L-arabinose	+	+	-
D-arabitol	+	+	+
cellobiose	+	-	-
i-erytritol	+	-	-
D-fructose	+	+	+
D-galactose	+	+	+
gentiobiose	+	-	-
α-D-glucose	+	+	+
m-inositol	+	-	-
α-D-lactose	v	-	+
lactulose	v	-	+
maltose	+	v	-
D-mannitol	+	+	+
D-mannose	v	+	+
D-melibiose	-	-	+
ß-methyl-D-glucoside	v	-	-
D-psicose	v	v	+
D-raffinose	-	-	+
D-sorbitol	+	+	+
sucrose	+	-	+
D-trehalose	+	-	+
furanose	+	-	-
methyl-pyruvate	+	+	+
mono-methyl-succinate	+	+	+
acetic acid	+	+	+
cis-aconitic acid	+	+	+
citric acid	+	v	v
formic acid	+	-	v
D-galactonic acid lactone	-	v	-
D-galacturonic acid	-	+	-
D-gluconic acid	+	+	+
D-glucosaminic acid	v	-	+
D-glucuronic acid	+	+	-
α-hydroxybutyric acid	-	-	+

ß-hydroxybutyric acid	+	+	+
γ-hydroxybutyric acid	-	v	v
α-ketobutyric acid	-	-	+
α-ketoglutaric acid	-	-	+
D,L-lactic acid	+	+	v
malonic acid	+	+	-
propionic acid	+	+	+
quinic acid	+	-	+
D-saccharic acid	+	-	+
succinic acid	+	+	+
bromosuccinic acid	+	+	+
succinamic acid	+	+	-
glucuronamide	-	v	-
D-alanine	+	v	-
L-alanine	+	+	-
L-alanyl-glycine	-	v	-
L-asparagine	+	+	v
aspartic acid	+	v	-
L-glutamic acid	+	+	+
L-histidine	v	-	-
hydroxy-L-proline	v	-	-
L-leucine	v	+	-
L-phenylalanine	v	-	-
L-proline	+	+	v
L-pyroglutamic acid	-	-	v
γ-amino butyric acid	+	-	-
urocanic acid	+	+	-
thymidine	+	+	-
putrescine	v	-	-
2-aminoethanol	+	+	-
2,3-butanediol	-	v	v
glycerol	+	+	+

The following substrates were not metabolized by all strains from cluster I, II and III: α-cyclodextrin, N-acetyl-D-galactosamine, L-fucose, xylitol, L-rhamnose, itaconic acid, p-hydroxyphenylacetic acid, α-ketovaleric acid, sebacic acid, alaninamide, glycyl-L-aspartic acid, glycyl-L-glutamic acid, L-ornithine, D-serine, L-serine, L-threonine, D,L-carnitine, inosine, uridine, phenylethylamine, D,L-α-glycerol-phosphate, glucose-1-phosphate, glucose-6-phosphate.

4. Discussion

Almost all seawater bacteria were recovered on the glass fibre filter in the "double Petri dish", and only weak bacterial growth was detected on nylon filter. Growing these bacteria from nylon filters, presumably ultramicrobacteria, in oligotrophic medium or natural seawater medium would be very time consuming, but can lead to isolations of ultramicrobacteria strains (Button et al., 1993; Schut

168

et al., 1993). Another dilution technique for enumeration and enrichment of oligotrophs has been described recently by Tan *et al.* (1996).

For marine bacteria preferring higher level of salt, Rüger and Krambeck (1994) used 23.5 g NaCl and 10.6 g MgCl$_2$.6H$_2$O dissolved in 1000 ml of deionized water for suspending the cells. This NaCl-MgCl$_2$ solution is not a buffer system and the pH after sterilization was 5.82. Despite this low pH, it was found that Biolog substrate utilization tests gave results not always identical with the utilization of organic substrates as sole carbon and energy source for growth. The authors therefore recommended to use the term substrate metabolism instead of substrate utilization for the Biolog identification system. Tan and Rüger (submitted) found that suspending the cells in NaCl-MgCl$_2$ solution of pH 7.0 resulted in oxidation of many more substrates to form the purple formazan by Antarctic oligotrophs belonging to cluster III. When cells were suspended in NaCl-MgCl$_2$ with a pH of 5.82, only tween 40, tween 80, mono-methyl-succinate, acetic acid, ß-hydroxybutyric acid, γ-hydroxybutyric acid, propionic acid, 2,3-butanediol and glycerol wells became purple (compare with Table 2). All Arctic strains showed positive reactions in Biolog GN MicroPlates. But many Antarctic strains belonging to cluster VI only showed few or not any reactions in Biolog GN MicroPlates, although the cells were suspended in NaCl-MgCl$_2$ solution with pH of 7.0. Marine bacteria probably prefer higher pH values than 7.0, since seawater has a pH of 7.6 to 8.0. This has to be investigated in further studies, by suspending the cells in buffer solution. Biolog, Inc. recommended Redox Purple, 200 µmol, which is a less toxic redox dye, for studies with oligotrophic bacteria (Bochner, personal communication).

The following 25 substrates were metabolized by all 74 psychrotrophic oligotrophs from cluster I, II and III: Tween 40, tween 80, D-arabitol, D-fructose, D-galactose, a-D-glucose, D-mannitol, D-mannose, D-psicose, D-sorbitol, methyl-pyruvate, mono-methyl-succinate, acetic acid, cis-aconitic acid, citric acid, D-gluconic acid, ß-hydroxybutyric acid, D,L-lactic acid, propionic acid, succinic acid, bromosuccinic acid, L-asparagine, L-glutamic acid, L-proline, and glycerol (Table 2). Substrates not utilized are listed in the footnote. Utilizations of 23 organic substrates by 106 psychrotrophic copiotrophs from Fram Strait and Western Greenland Sea (expedition ARKTIS IV/2, 1987) were tested at 4 and 20°C by Tan and Rüger (1991). Most carbohydrates, organic acids, alcohols, and L-alanine were assimilated at both temperatures, but arginine, aspartate and ornithine were used only at 20°C by many strains.

Staley (1993) wrote in his review article that N-acetyl-D-glucosamine (NAG) is a hydrolysis product of chitin, a substance found in the exoskeleton of krill and other zooplankton species. Biomass of krill has been estimated to amount from 0.5 to 3 billion metric tons, wet weight. Chitin comprises from 4 to 10% of the dry weight, and krill consist of 76 to 80% moisture. Therefore, about 0.16 to 75 million metric tons of chitin arise from this source alone. Interestingly, NAG was metabolized by strains from cluster III, originated from Arctic and Antarctic regions. Strains belonging to cluster I and V were able to metabolize cellobiose, a degradation product of cellulose. As mentioned by Wynn-Williams (1990), ice algae among the sea ice microbial communities are known to tolerate low temperatures and high salinities by accumulating polyols. Tween 40 and 80 are polyols. Both substrates and also glycerol were metabolized by strains from clusters I, II, III and IV.

Biolog metabolic fingerprints of Antarctic oligotrophs belonging to cluster IV, V and VI will be reported in Tan and Rüger (submitted). The 9 strains from cluster IV could metabolize glycyl-L-glutamic acid, L-serine, L-threonine and inosine, which were not utilized by the other isolates. For these strains, Biolog substrate utilization tests were performed at 12° C, which is not the optimum temperature for growth, but viability of stock cultures kept at this temperature was better than at 2°C. Only strains from cluster V were able to metabolize α-cyclodextrin, N-acetyl-D-galactosamine and ß-methyl-D-glucoside. The latter was also metabolized by strains from cluster IV. The Antarctic isolates did not utilize methylamine as sole carbon and energy source.

These oligotrophic bacteria from north and south polar regions can be grown on Basal Seawater Agar and colonies were visible after 1 week of incubation at 12 or 20°C. Therefore, these psychrotrophic oligotrophs will be the subject of investigations in studying growth rates at low-nutrient concentrations and low temperatures. Phylogenic studies with the same groups of bacteria from the north and south polar regions can give informations about the strategies of cold adaptations in bacteria.

5. Acknowledgements

The technical assistance of Annegret Mädler and Karin Springer is greatly acknowledged. I am obliged to H-J. Rüger for helping me, running the cluster programme on PC, and to the reviewers for their critical comments. This research was supported partly by a grant from the German Research Council (Ta 63/7-1).

This is contribution no. 1144 of the Alfred-Wegener-Institute for Polar and Marine Research, Bremerhaven.

6. References

Button DK, Schut F, Quang P, Martin R, Robertson BR (1993) Viability and isolation of marine bacteria by dilution culture: Theory, procedures, and initial results. Appl Environ Microbiol 59: 881-891

Carlucci AF, Shimp SL, Craven DB (1986) Growth characteristics of low-nutrient bacteria from the north-east and central Pacific Ocean. FEMS Microbiol Ecol 38: 1-10

Fütterer DK, Schrems O (eds) (1991) The Expedition ANTARKTIS-VIII of RV "Polarstern" 1989/90. Reports of Legs ANT-VIII/6-7. Rep. Pol. Res. 90: 231 pp Buchhandlung Karl Kamloth, Bremen

Ishida Y, Eguchi M, Kadota H (1986) Existence of obligately oligotrophic bacteria as a dominant population in the South China Sea and the West Pacific Ocean Mar Ecol Prog Ser 30: 197-203

Krambeck H-J, Witzel K-P (1983) Classification of aquatic bacterial strains: An example of numerical taxonomy in limnology. EDV in Medizin und Biologie 14: 45-49

Krause G, Meincke J, Thiede J (eds) (1989) Scientific Cruise Reports of Arctic Expeditions ARK IV/1, 2 & 3. Rep. Pol. Res. 56: 146 pp Buchhandlung Karl Kamloth, Bremen

Rüger H-J, Krambeck H-J (1994) Evaluation of the BIOLOG substrate metabolism system for classification of marine bacteria. System Appl Microbiol 17: 281-288

Schut F, de Vries EJ, Gottschal JC, Robertson BR, Harder W, Prins RA, Button DK (1993) Isolation of typical marine bacteria by dilution culture: Growth, maintenance, and characteristics of isolates under laboratory conditions. Appl Environ Microbiol 59: 2150-2160

Sneath PHA, Sokal RR (1973) Numerical Taxonomy. The principles and practice of numerical classification, Freeman WH and company, San Francisco

Staley JT, Herwig RP (1993) Degradation of particulate organic material in the Antarctic. In: Friedmann EI, Thistle AB (eds) Antarctic Microbiology, Wiley-Liss, New York, pp 241-264

Tan TL (1986) Construction and use of a dialysis chamber for investigating in situ the toxicity of heavy metals on bacteria. IFREMER, Actes de Colloques 3: 589-595

Tan TL, Rüger H-J (1991) Biomass and nutritional requirements of psychrotrophic bacterial communities in Fram Strait and Western Greenland Sea. Kieler Meeresforsch Sonderh 8: 219-224

Tan TL, Reinke M, Rüger H-J (1996) New dilution method in microtiter-plates for enumeration and enrichment of copiotrophic and oligotrophic bacteria. Arch Hydrobiol 137: 511-521

Tan TL, Rüger H-J (submitted) Enrichment, isolation, and BIOLOG metabolic fingerprints of oligotrophic bacteria from the Antarctic Ocean. System Appl Microbiol

Wynn-Williams DD (1990) Ecological Aspects of Antarctic Microbiology. In: Marshall KC (ed) Advances in Microbial Ecology, Volume 11, Plenum Press, New York, pp 71-146

The Influence of Microbial Community Structure and Function on Community-Level Physiological Profiles.

Jay L. Garland[1], K.L. Cook[1], C. A. Loader[1], and B. A. Hungate[2]

[1] Dynamac Corporation, Mail Code DYN-3, Kennedy Space Center, Fla. 32899, USA
[2] Smithsonian Environmental Research Center, PO Box 28, Edgewater, MD, 21037, USA

Abstract. Patterns of carbon source utilization, or community-level physiological profiles (CLPP), produced from direct incubation of environmental samples in BIOLOG microplates can consistently discriminate spatial and temporal gradients within microbial communities. While the resolving power of the assay appears significant, the basis for the differences in the patterns of sole carbon source utilization among communities remains unclear. Carbon source utilization as measured in this assay is a measure of functional potential, rather than *in situ* activity, since enrichment occurs over the course of incubation, which can range from 24 to 72 hours (or even longer) depending on inoculum density. The functional profile of a community could be an indicator of carbon source availability and concomitant selection for specific functional types of organisms. A more limited view of the profile is as a composite descriptor of the microbial community composition without any ecologically relevant functional information. We manipulated microbial community structure and function in laboratory microcosms to evaluate their influence on CLPP. The structure of rhizosphere communities was controlled by inoculating axenic plants (wheat and potato) with different mixed species (non-gnotobiotic) inocula. Inoculum source influenced CLPP more strongly than plant type, indicating that CLPP primarily reflected differences in microbial community structure than function. In order to more specifically examine the influence of microbial function on CLPP, specific carbon sources in the BIOLOG plates (asparagine and acetate) were added to a continuously stirred tank reactor (CSTR) containing a mixed community of microorganisms degrading plant material. Daily additions of these carbon sources at levels up to 50% of the total respired carbon in the bioreactor caused significant changes in overall CLPP, but caused no, or minor, increases in the specific response of these substrates in the plates. These studies indicate that the functional relevance of CLPP should be interpreted with caution.

Keywords. Communities, BIOLOG, carbon sources, rhizosphere, bioreactor

1. Introduction

Profiles of the potential utilization of multiple sole carbon sources by mixed microbial communities can be rapidly produced through direct inoculation of environmental samples (e.g., water, bacterial suspensions from soil or biofilms) into Biolog microplates (Garland and Mills 1991). Utilization of the 95 different sole carbon sources in the separate wells of the microplate is quantified by measuring the reduction of tetrazolium violet to colored formazan in respiring cells. The patterns of carbon source utilization, or community-level physiological profiles (CLPP), can consistently discriminate spatial and temporal gradients within microbial communities (Garland and Mills 1991, Garland and Mills 1994, Lehman *et al.* 1995, Winding 1994). Relatively subtle shifts in microbial communities, such as the response of rhizosphere communities to plant age (Garland 1996b) are discernible with the method. Discrimination is enhanced if differences in inoculum density and the concomitant effects on the overall rate of color formation are accounted for either in the inoculation of the plates or data analysis (Garland 1996a, Haack *et al.* 1995).

While the resolving power of the assay appears significant, the basis for the differences in the patterns of sole carbon source utilization among communities requires further research. Carbon source utilization as measured in this assay is a measure of functional potential, rather than *in situ* activity, since enrichment occurs over the course of incubation, which can range from 24 to 72 hours (or even longer) depending on inoculum density (Garland, 1996a). The ecological relevance of these changes in functional potential remains unclear. Do CLPP reflect *in situ* carbon source availability and concomitant shifts in the relative abundance of organisms capable of utilizing these compounds? The presence or absence of response in a specific well in the community-level assay does appear to reflect the presence or absence of individuals in the community capable of utilizing the substrate. Haack *et al.* (1995), working with simple gnotobiotic mixtures of bacteria (4 or 6 different strains), reported that a positive response reflected the presence of specific strains capable of using that substrate, while a negative response typically indicated the absence of individuals capable of degrading that substrate. Lehman *et al.* (1995) found that community-level profiles of deep subsurface communities rarely showed a positive response toward any carbohydrates, and a similar lack of utilization of sugars has been reported for isolates from deep subsurface samples (Fredrickson *et al.* 1991).

However, consistent differences in the community-level profiles among samples from complex, natural communities (e.g., soil types, rhizosphere communities from different plant types) are typically the result of variation in the rate and extent, rather than the presence or absence, of responses in different wells. Experience in our lab has indicated that most microbial samples with cell densities greater than 10^5 ml^{-1} show positive responses toward the vast majority of substrates after extended incubations, but that the relative pattern of utilization at specific points during the incubation can be used to consistently discriminate sample types. Garland and Mills (1991) hypothesized that the degree of response may reflect the relative abundance of organisms able to utilize a particular substrate. This conceptual view of community-level physiological profiles as guild profiles did not fit results from studies with simple gnotobiotic mixtures. Haack *et al.* (1995) found that although different mixtures of bacterial strains produced distinctive profiles of carbon source utilization, the rate of response in individual wells was not correlated to the number of organisms in the inoculum capable of utilizing the substrate. The relationship between CLPP and the distribution of functional abilities within natural, complex communities has not been established. Despite the lack of a clear linkage, CLPP are being used as indicators of *in situ* carbon source utilization (Insam *et al.* 1996) and functional biodiversity (Zak *et al.* 1994).

We experimentally manipulated microbial community structure and function in laboratory microcosms to determine their affects on CLPP. Microbial communities with distinctive community structures, but similar functional attributes, were created by introducing distinctive mixed microbial inocula into axenic plants. This rhizosphere inoculation experiment addressed the question: Do structurally dissimilar but functionally similar communities yield dissimilar CLPP? Community function was altered by amending aerobic bioreactors with specific carbon sources (i.e., asparagine and acetate) found in the Biolog microplates. These bioreactor studies addressed the question: Do shifts in CLPP correspond to known changes in carbon source availability?

2. Material and Methods

2.1 Rhizosphere inoculation study

Wheat seeds (*Triticum aestivum* L. cv. Yecora roja) were sterilized using mercuric chloride and hydroxylamine hydrochloride (Barber 1967) and germinated for five days on moistened filter paper. Sterility of the plants was confirmed by growing randomly selected seedlings in 50 ml culture tubes containing 15 ml 0.55% R2A agar (Difco Laboratories) and checking for lack of visible microbial growth in the rhizosphere. Axenic white potato (*Solanum tuberosum* L. cv. Norland) plants were produced from nodal cuttings (Hussey and Stacey 1981), and grown for 4 weeks in Murashige and Skoog's media (Murashige and Skoog 1962) to allow for root development (Hussey and Stacey 1981). Sterility was confirmed by the lack of visible microbial growth in the rhizosphere of the sucrose-containing media.

Potato and wheat plants were placed into autoclaved foam plugs and transferred into sterile wide-mouth glass vessels containing 50 ml of 1/4 strength Hoagland's solution (Morales *et al.* 1996). These are subsequently refered to as plant growth vessels. Plants were grown for seven days on a bench top plant growth system containing fluorescent lamps and a humidifier enclosed within black plastic. Plants growth vessels were placed on a rotating shaker table (75 rpm) to provide aeration. Environmental conditions were as follows: 12h/12h light/dark, average photosynthetic photon flux of 100 μmoles m^{-2} sec^{-1}, continuous temperature at 22±1° C, and 50±5% relative humidity.

Plant growth vessels were inoculated with mixed microbial inocula from the rhizosphere of mature wheat and potato plants grown in recirculating hydroponic systems. Separate inocula were prepared by shaking excised root samples from wheat and potato, respectively, in hydroponic solution, growing the resulting suspension for 24 h in R2A broth, and freezing at -70°C in glycerol. In preparation for plant growth studies, the frozen vials of each inoculum type were thawed and 20 μl was added to 50 mls of R2A media. Three separate cultures were initiated; 1) wheat inoculum alone, 2) potato inoculum alone, and 3) wheat and potato inocula combined. After 16 h growth with shaking at 200 rpm, the cultures were centrifuged at 2000 g for 20 min, and the pellet was resuspended in 1/4 strength Hoagland's solution to a density of 0.1 abs. units at 600 nm. These suspensions (0.5 ml) were added to the nutrient solution of plant growth vessels immediately before addition of the plant. Four replicate microcosms containing each plant type were inoculated with each inoculum type (4 reps x 2 plant types x 3 inocula = 24 total plants).

After 7 days growth, plants were removed from the vessels, and roots were excised and placed into 0.85% saline. Suspensions of rhizosphere microorganisms were produced by shaking the roots in glass beads as previously described (Garland, 1996b). The suspensions were diluted 10-fold in 0.85% saline, and inoculated into Biolog GN microplates. Samples of the suspensions also were serially diluted on R2A agar for estimation of culturable bacteria and fixed for subsequent estimation of total numbers using the acridine-orange (AO) method with epiflourescent microscopy (Hobbie *et al.* 1977). Plant growth was estimated by measuring shoot length (wheat), shoot dry weight (potato), and root weight (both).

2.2 Bioreactor studies

An 8L continuously-stirred tank reactor (CSTR) containing dried plant material (plant material (25 g/L) was used for these studies (Finger and Strayer 1994). Relevant operating conditions were as follows: 35 ° C temperature, 6.5 pH, and 7.5 L min^{-1} air flow rate, and 300 rpm stirring rate. Nominal operation of the reactor involved daily addition of 25 g of milled (2 mm) oven-dried wheat biomass (leaves, stems, and roots) and 1 L of deionized water following removal of 1 L of reactor contents. The ground biomass was added gradually over the course of the day using a screw-feed mechanism. Bioreactor headspace gas was circulated through a carbon dioxide analyzer (Model LI-6251, LI-COR). Approximately 40-50 % of the volatile solids of the wheat biomass were degraded with this 8 day retention time.

After nearly 3 months of continuous operation in the nominal mode, specific carbon amendment studies were performed. The first experiment involved daily additions of 1.5 g of asparagine for 10 consecutive days, followed by addition of 5 g for another 9 days. The asparagine was dissolved in 100 mls of deionized water, all of which was added to the reactor immediately after the daily harvesting/feeding schedule. Samples were taken from the daily harvests of the bioreactor, diluted 20 fold with 0.85% NaCl, and blended for 30 sec. The resulting suspension was diluted 20 fold in 0.85% NaCl, and inoculated into Biolog GN microplates.

After 21 days of nominal operation following the asparagine amendment study, daily amendments with 5 g of acetate were begun. The reactor was amended for 8 straight days, then amended for 2 additional days after being left unamended for 2 days. The reactor was sampled as above.

2.3 Data analysis

2.3.1 Plate reading

Plates were read over a time course of incubation as previously reported (Garland 1996a,b). Comparisons were made among plate readings of equivalent average well color development (AWCD).

2.3.2 Statistical analysis

Differences in CLPP among rhizosphere samples were tested statistically by comparing principal component scores among treatment groups using analysis of variance. Principal component scores 1 and 2 were each analyzed using a two-way analysis of variance (ANOVA); the two factors were inoculum source (3 levels, potato, wheat, and potato+wheat) and plant type (2 levels, potato and wheat). Tukey's post hoc tests were used to determine significant differences ($p<0.05$) between individual pairs of means. Plant growth data were analyzed using a one-way ANOVA with inoculum as the factor.

We used one-way ANOVAs to test for differences in principal component scores and the response of the amended carbon source. Mean values for each sampling time with common levels of asparagine (0 g, 1 g, 5g) and acetate (0 g or 5 g) amendments were used as replicates for the asparagine and acetate amendment "treatments". Because we observed a delayed response to the acetate addition and had difficulties with pH control immediately after acetate addition (see Figure 3), we excluded the first three time points after acetate addition from the 5 g treatment group. Data were log-transformed as necessary to conform with the ANOVA assumption of homogeneity of variance.

3. Results

3.1 Seedling inoculation experiment

PCA revealed distinctive CLPP between the different rhizosphere samples. The first principal component (PC 1), which reflects the primary effect in the data (explained 38.7% of the variance), separated rhizosphere samples based on inoculum source (Fig. 1). The wheat and potato inocula produced the most divergent PC 1 scores, and the mixed wheat-potato inoculum yielded

176

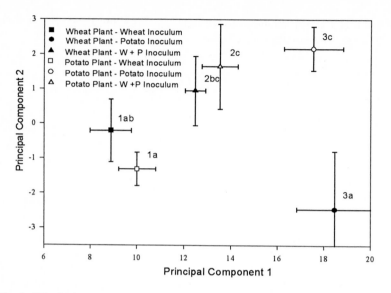

Fig. 1. Principal component analysis of rhizosphere inoculation study. Points represent means and standard deviations of 4 replicate samples. Means with different numbers have significantly different scores for the first principal component. Means with different letters have significantly different scores for the second principal component.

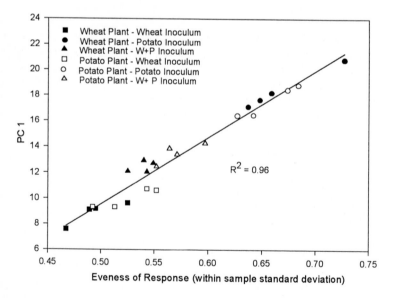

Fig. 2. Correlation of PC1 scores and evenness in response (as expressed as within samples standard deviation) for samples from rhizosphere inoculation study.

Table 1. Correlation of carbon sources to the first two principal component of the data (0.75 AWCD) from the seedling inoculation experiment

PC 1		PC 2	
C source	r^1	C course	r
cis-aconitic acid	0.846	N-acetyl-D-galactosamine	0.898
quinic acid	0.863	gentiobiose	0.888
L-asparagine	0.842	maltose	0.822
L-aspartic acid	0.779	D-sorbitol	0.875
L-pyroglutamic acid	0.857	inosine	0.765
dextrin	-0.927	bromo-succinic acid	-0.721
tween 80	-0.853	L-histidine	-0.714
D-galactose	-0.844	L-proline	-0.646
m-inositol	-0.833		
lactulose	-0.812		
alpha-ketovaleric acid	-0.874		
sebacic acid	-0.892		
alaninamide	-0.835		
L-alanine	-0.864		
L-alanyl glycine	-0.909		
L-leucine	-0.826		
L-ornithine	-0.867		
L-phenyalanaine	-0.858		
urocanic acid	-0.885		

[1] Pearson's correlation coefficient

intermediate scores. Plant type did not significantly affect PC 1 (p=0.33), whereas inoculum source did (p=0.004). The second principal component (PC 2), which reflects the secondary effect in the data (explained 24.8% of the variance) reflected a difference between plant type, but only for plants receiving the potato inoculum (Fig. 1). Accordingly, plant type (p=0.004), inoculum source (p=0.003), and the interaction between the two factors (p <0.001) were all significantly affected by PC 2.

None of the measured plant growth parameters, including wheat root dry wt (p=0.768), potato root dry wt (p=0.420), wheat shoot length (p=0.591), and potato shoot dry wt (p=0.934), were significantly affected by inoculation. Thus, with respect to rhizosphere factors affected by plant growth, the rhizospheres of the plants receiving different inocula were functionally similar.

The specific carbon sources responsible for the differences in CLPP among samples are reported in Table 1. More carbon sources showed a strong negative rather than a strong positive correlation to PC 1. Since all samples were of similar overall response (i.e., AWCD of 0.75), the cause of this disproportionate response may not be intuitively obvious. However, the average response for the positively correlated substrates ranged from 1.5 to 1.8 abs. units, while the response of negatively correlated substrates ranged from 0.15 to 0.60 abs. units (data not shown). Since a similar proportional change in a more strongly responsive well would result in a larger absolute changes in abs. units, a sample with increases in a few strongly responsive wells and decreases in a greater number of less responsive wells would still yield the same overall AWCD. This suggests that PC1 might reflect differences among samples in the relative evenness in the CLPP. The standard deviation in the response among wells for a given sample is inversely related to the evenness of well response in the plate. As predicted, the standard deviation of samples was highly correlated ($R^2 = 0.96$) to their PC 1 scores (Fig. 2).

3.2 Bioreactor studies

3.2.1 Asparagine amendment

The added asparagine was rapidly respired as reflected by the spike in the concentration of CO_2 in the headspace immediately after feeding. (Fig. 3a). Both the height and duration of the spike increased when the amendment level was changed from 1.25 to 5.0 g/day. During normal operation, approximately 1.8-2.8 g CO_2 day^{-1} are produced from the partial degradation of the plant biomass. Addition of 1.25 g and 5.00 g of asparagine increased CO_2 production by 13-20% and 39-50%, respectively.

Asparagine addition decreased PC1 (which accounted for 24% of the variance in the data), and the decrease was proportional to the amount of asparagine added (ANOVA and multiple range tests, $p<0.05$) (Fig. 3b). However, addition of asparagine had no effect on color development in the asparagine well of the BIOLOG plates ($p=0.66$) (Fig. 3c). Several other carbon sources were responsible for the shift in the CLPP, as indicated by their correlation to PC 1 (Table 2).

Results from the analysis of samples at an AWCD of 0.50 are reported here since asparagine response was intermediate (i.e., 0.6-0.8 abs. units) at that point in overall color development. However, the same effects, both in terms of PC 1

Table 2. Correlation of carbon sources to the first principal component of the data (0.50 AWCD) from the asparagine amendment study

Increased Response w/Amendment		Decreased Response w/Amendment	
Carbon source	r^1	Carbon Source	r^1
alpha-ketovaleric acid	-0.762	adonitol	0.829
succinamic acid	-0.778	fucose	0.834
glucuronamide	-0.781	D-raffinose	0.826
L-leucine	-0.693	D-sorbitol	0.701
2,3-butanediol	-0.762	D-arabitol	0.725
		D-glucuronic acid	0.734
		itaconic acid	0.790
Amended substrate			
L-asparagine	-0.486		

[1] Pearson's correlation coefficient

scores and specific asparagine response, were apparent for analysis of data at AWCD of 0.25 (when asparagine was beginning to respond), and AWCD of 0.75 (when asparagine response was becoming saturated) (data not shown).

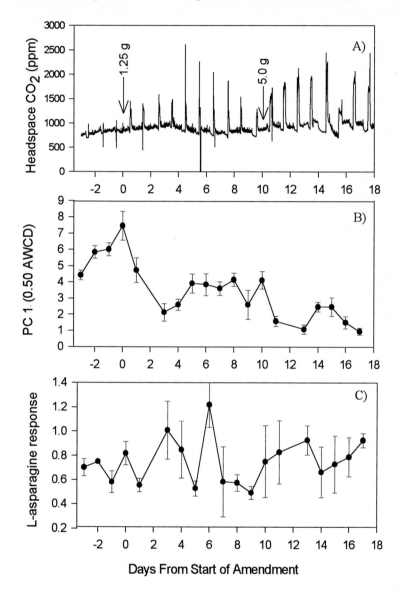

Fig. 3. Effects of asparagine amendment on A) headspace CO_2 concentration (5 min averages), B) PC 1 scores from analysis of community-level physiological profiles (mean and standard deviation of 3 samples, and C) color response in asparagine well (mean and standard deviation of three samples).

3.2.2 Acetate amendment

As observed with asparagine, the addition of acetate resulted in a pulse of respiration (Fig. 4a). Addition of 5.00 g of potassium acetate increased CO_2 production by 30-40% .

A shift in CLPP with acetate amendment was apparent from analysis of the first PC, which accounted for 25% of the total variance in the data. The shift did not occur until the fourth day after amendment (Fig. 4b), as reflected by the significant effect of amendment on the PC 1 scores if the first three days after amendment are excluded ($p<0.01$), and the lack of significance if all data are included ($p=0.16$). The sudden shift in CLPP corresponded to a period immediately after the increase in pH.

In contrast to the trend observed with asparagine, the acetate response closely followed changes in PC 1 ($r =0.602$). As observed with the PC 1 scores, amendment had a significant effect on acetate response if the first three days after amendment are excluded ($p<0.01$), and the lack of significance if all data are included ($p=0.12$). It is important to note that even after amendment, acetate response was relatively weak (i.e., < 0.50 abs. units in a plate with an average well response of 1.00).

Results from the analysis of samples at an AWCD of 1.00 are presented since acetate response was below detectable levels at earlier set points in color development. Similar trends in the PC 1 scores were observed when samples at 0.25, 0.50, and 0.75 were analyzed, although acetate response was not correlated to the axis because it was minimal (data not shown).

Table 3. Correlation of carbon sources to the first principal component of the data (1.00 AWCD) from the acetate amendment study

Increased Response w/Amendment		Decreased Response w/Amendment	
Carbon source	r^1	Carbon Source	r^1
D-fructose	0.735	D-alanine	-0.649
alpha-D-glucose	0.656	L-alanine	-0.653
D-mannitol	0.642	L-asparagine	-0.677
citric acid	0.712	L-glutamic acid	-0.794
glycyl-L-glutamic acid	0.690	L-proline	-0.786
		L-pyroglutamic acid	-0.684
		gamma-amnio butyric acid	-0.682
		phenylethylamine	-0.853
Amended substrate			
acetate	0.603		

[1] Pearson's correlation coefficient

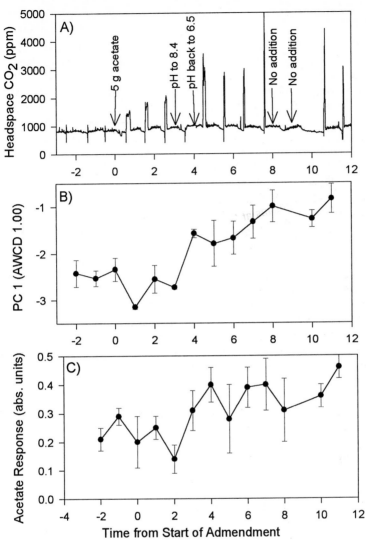

Fig. 4. Effects of aectate amendment on A) headspace CO_2 concentration (5 min averages), B) PC 1 scores from analysis of community-level physiological profiles (mean and standard deviation of 3 samples, and C) color response in acetate well (mean and standard deviation of three samples).

4. Discussion

Results from this study indicate that the functional relevance of CLPP should be interpreted with caution. This conclusion is based on two lines of evidence. First, changes in CLPP were produced when only community structure was varied (i.e., rhizosphere inoculation experiments. Secondly, shifts in response to *in situ* carbon source availability involved weak or no change in the response of the amended substrate, even when it represented up to 50% of the readily available carbon.

The significant increase in acetate response, a weakly responsive test, in the carbon amendment study suggests that the presence/absence of a response in the CLPP may provide information on carbon source availability. However, the relative rate of utilization does not appear to provide useful functional information. This conclusion is based on the fact the response of asparagine did not increase with amendment, and the fact that neither acetate nor asparagine became the dominant response in the assay, even though they represented from 30-50% of the total respired C *in situ*. These results with complex microbial communities concur with previous studies with gnotobiotic mixtures (Haack *et al.* 1995).

An ancillary finding from the rhizosphere inoculation study was that the evenness in response can significantly influence classification of CLPP. This finding is relevant because the change in evenness of response was the result of variation in community structure, not function. It is unclear from this study if the change was due to the diversity of organisms present, and/or the metabolic diversity of individual organisms. However, it is clear that effects of community structure on the evenness of response will confound interpretation of functional diversity (Zak *et al.* 1994). Thus, richness - in this case, the number of positively responding Biolog wells - may be more functionally relevant than evenness.

Further carbon amendment studies could improve our understanding of CLPP. Amendments to more carbon-limited communities may be useful to mimic conditions in many natural environments. Preliminary findings from this study suggest that the interaction between carbon source availability and physiochemical stress (i.e., pH) should be evaluated. The delayed response to acetate amendment corresponded to a transient pH increase. One interpretation of the results is that the bioreactor community did not respond structurally to the supplemental carbon because it possessed an assimilative capacity, but that the pH stress restructured the community. Further experiments are necessary, but CLPP may be useful tool for evaluating stability in microbial communities.

This work has demonstrated apparent limitations of CLPP for measuring functional attributes of microbial communities. At the same time, this research has demonstrated that CLPP can reproducibly discriminate between microbial communities in different experimental treatments. Our results indicate that CLPP describe microbial community structure, but further work is needed to define what components of the *in situ* microbial community CLPP describe, as well as whether CLPP, perhaps in combination with other techniques, might provide any taxonomic information.

5. References

Barber DA (1967) The effects of microorganisms on the absorption of inorganic nutrients by intact plants. J Exp Bot 18:163-169

Finger BW, Strayer RF (1994) Development of an intermediate-scale aerobic bioreactor to regenerate nutrients from inedible crop residues. SAE Technical Paper 941501

Fredickson JK, Balkwill DL, Zachara JM, Li SW, Brockman FJ, Simmons MA (1991) Physiological diversity and distributions of heterotrophic bacteria in deep Cretaceous sediments of the Atlantic coastal plain. Appl Environ Microb 57: 402-411

Garland JL (1996a) Analytical approaches to the characterization of samples of microbial communities using patterns of potential carbon source utilization. Soil Biol Biochem 28:213-221 .

Garland JL (1996b) Patterns of potential carbon source utilization by rhizosphere communities. Soil Biol Biochem 28:223-230

Garland JL, Mills AL (1991) Classification and characterization of heterotrophic microbial communities on the basis of patterns of community-level sole-carbon-source utilization. Appl Envrion Microb 57:2351-2359

Garland JL, Mills AL (1994). A community-level physiological approach for studying microbial communities, pp. 77-83. In Ritz K, Dighton J, Giller KE (ed.) Beyond the biomass: compositional and functional analysis of soil microbial communities. John Wiley & Sons, Chichester, UK

Haack SK, Garchow H, Klugg MJ, Forney LJ (1995) Analysis of factors affecting the accuracy, reproducibility, and interpretation of microbial community carbon source utilization profiles. Appl Environ Microb 61:1458-1468

Hobbie JE, Daley RJ, Jasper S (1977) Use of Nucleopore filters for counting bacteria for fluorescent microscopy. Appl Environ Microb 33:1225-1228

Hussey G, Stacey NJ (1981) In Vitro propagation of potato (*Solanun tuberosum* L.) Annals Bot 48:787-796

Insam H, Amor K, Renner M, Crepaz C (1996) Changes in functional abilities of the microbial community during composting of manure. Microb Ecol 31:77-87

Lehman RM, Colwell FS, Ringelberg DB, White DC (1995) Combined microbial community-level analyses for quality assurance of terrestial subsurface cores. J Microb Meth 22:263-281

Morales A, Garland JL, Lim DV (1996) Survival of potentially pathogenic human-associated bacteria in the rhizosphere of hydroponically-grown wheat. FEMS Microbiol Ecol 20:155-162

Murashige T, Skoog F (1962) A revised medium for rapid growth and bioassays with tobacco tissue cultures. Physiol Plant 15:473-497

Winding A (1994) Fingerprinting bacterial soil communities using Biolog microtitre plates, pp. 85-94. In Ritz K, Dighton J, Giller KE (eds.) Beyond the biomass: compositional and functional analysis of soil microbial communities. John Wiley & Sons, Chichester, UK

Zak JC, Willig MR, Moorehead DL, Wildman HG (1994) Functional diversity of microbial communities: a quantitative approach. Soil Biol Biochem 26:1101-1108

Strain and Function Stability in Gnotobiotic Reactors

Aaron L. Mills[1] and Judith E. Bouma[1,2]

[1] Laboratory of Microbial Ecology, Department of Environmental Sciences, University of Virginia, Charlottesville, VA 22903, USA
[2] Present Address: NASA, Mail Code SD-4, NASA-JSC, Houston, TX 77058, USA

Abstract. The ability of the community level physiological profiling approach to analyzing microbial community structure was examined in a constructed community. Four bacterial strains, characterized by BIOLOG®-GN profiles, were combined into 2 communities which were maintained for 7 weeks on 2% wheat residue in distilled water with 10% daily removal and replenishment of medium. after 3 weeks incubation, aeration was stopped for 5 days. Although the abundance of all the isolates decreased during the perturbation, all the strains persisted throughout the experiments. Only 80% of functions lost during perturbation were recovered afterward. Persistence of strains in a community did not guarantee persistence of metabolic functions which those strains could perform.

The approach to community structure showed that simple summation of pure culture functional profiles did not accurately predict the functional profiles of the mixed communities. Functions resulting from synergistic interaction between the strains appeared in the mixed communities, but were inactive during the perturbation. The assay plates could not detect metabolic activities of strains present at densities below $\approx 1 \times 10^7$ CFU ml^{-1}. Population densities of component strains in the community could not be inferred from community-level functional profiles. Because strains behaved differently in mixed and pure cultures, community behavior was not accurately predicted by pure culture performance of even the densest strains in the communities.

Keywords. Bacteria, community structure, stability, resiliency

1. Introduction

The recognition that the behavior of communities of microorganisms in the environment is not well predicted by studies of pure cultures in laboratory situations has led to a flurry of effort to detail the structure and function of the communities and their constituents. Studies fall into two broad categories: one includes those investigations concerning the genetic or taxonomic structure of the community, and the second focuses on the functional characteristics of the community and its members. The use of molecular techniques offers insight into the genetic composition of the organisms that constitute the community, and the recent efforts to extract RNA and DNA from whole communities extend that insight to the intact community. Some of the molecular techniques also have promise for examining functional aspects of the microbial communities, especially if probes for loci that code for enzymes of importance are available. The use of the physiological profiles, specifically, those dealing with sole carbon source utilization (Garland, 1996; Garland and Mills, 1991; Insam et al., 1996; Winding, 1994; Zak et al., 1994), is one of several techniques that should be considered as a means of judging similarity among microbial communities (Turco et al., 1994)

The use of physiological profiles offers another view of the functional abilities of microbial communities. Most often, the profiles are based on use of a series of organic substrates as sole sources of carbon, energy, and electrons. The use of commercially prepared test kits such as the BIOLOG® or PHENE-PLATE® assays can make the extensive testing necessary much easier as well as making the results more comparable from study to study. Questions arise, however, in using the commercial preparations with regard to reproducibility and interpretation.

We have conducted a study of a constructed communities in an effort to examine specifically, some of those questions. The use of a small number of isolates in a constructed community permitted the meaningful evaluation of the physiological profile approach to community structure in the context of determining if the assay can be used to examine population changes in a dynamic community, and if the approach produces reproducible results when applied to similar or identical samples. Furthermore, we sought to determine if the response patterns in the whole community reflected a summation of the properties of the individuals. Answers to each of these questions are essential to the use of the approach in microbial community ecology.

2. Methods and Materials

2.1 General approach

Detailed methods for isolation of the test strains, construction of the reactors, and conduct of the experiments are given in Bouma (1995) and Bouma and Mills (submitted). A brief description of the supporting work is presented here.

Four bacterial strains used as building blocks for the gnotobiotic community capable of degrading inedible plant material were isolated and functionally characterized. The strains were assigned numbers (6a, 9, 13a, and 19), and BIOLOG®-GN profiles were obtained, but no additional attempt was made to assign a conventional name to the organisms. For enumeration of separate population densities in mixed culture, strains with distinct colony morphologies were chosen. The strains were functionally diverse in their ability to degrade the different components of inedible (human) plant matter and had distinguishable functional profiles on BIOLOG®-GN assay plates (Bouma, 1995). Pure culture growth patterns, both population densities and functional activity profiles, were determined for comparison with observations from mixed culture.

The four chosen strains were combined into parallel gnotobiotic communities in sterile, aerated 3-liter reactor vessels. Mixed cultures were maintained for 6 to 7 weeks, with daily removal of one tenth of the culture volume and replenishment with an equal volume of fresh sterile medium containing inedible plant material. Every effort was made to prevent recruitment of new strains (contaminants) into the communities.

Population dynamics of the communities were determined by spread plate enumeration of viable cells on TSA plates at least three times a week. Functional assays if each community's ability to degrade the 95 different compounds on the BIOLOG®-GN plates were performed once a week. Data obtained from simultaneous monitoring of both population and functional dynamics through a perturbation (anoxia) allowed examination of the relationship between population structure within a community and functional behavior of that community.

2.2 Production of physiological profiles with BIOLOG®-GN plates

Since the goal of the physiological profiles of the mixed community was to monitor the functional activity of the mixed liquid culture, the standard BIOLOG® protocol had to be modified. An additional complication was that part of the organic substrate in the liquid medium in this study was present as particulate matter. For the BIOLOG® assay to be successful, all extraneous carbon sources must be removed from the bacterial inoculum added to the assay plate. The ability of an isolate to be cleanly separated from the particulate matter in 2% wheat liquid medium, resuspended in saline, and then to produce a BIOLOG® response in a timely fashion was the first hurdle for candidate members of the mixed community.

A uniform protocol was established for extracting cells for both pure and mixed culture functional profile determinations. A 100 to 150-ml sample of liquid culture was placed in a sterile 500-ml Erlenmeyer flask, shaken vigorously for 15-20 seconds, and allowed to settle for 2-4 min. Carefully avoiding the solid residue at the bottom of the flask, approximately 90 ml of liquid was removed to a sterile 250-ml flask, and allowed to settle for another 15 min. Equally-sized 35 to 37-mL aliquots of the cloudy golden brown liquid were pipetted into two sterile 50-ml polypropylene centrifuge tubes, again avoiding any solid residue settled on the bottom of the source flask. The tubes were capped and spun for 18-20 minutes at $27,100 \times g$. The clear brown supernatant was decanted and discarded. The cells forming the top layer of each pellet were resuspended in 10 to 15 ml of sterile 0.85% NaCl. The total volume of saline (20 to 30 ml) used to resuspend both pellets of a sample varied inversely with the suspected density of cells in the sample (density was estimated from the plate-count enumeration history of that strain or culture and the size of the cell layer in the pellet). The standard volume of resuspended cells was 30 ml.

Duplicate 100 µl samples of the resuspension were diluted through parallel series and spread onto TSA plates for enumeration of CFU ml^{-1}. Five replicate 1:

10 dilutions of the original resuspension were created by pipetting 2.5 ml of the original resuspension into 22.5 ml of sterile 0.85% saline. Each replicate was used to inoculate a BIOLOG®-GN plate with 150 µl of diluted resuspension per well. The assay plates were incubated in the dark for 24 h at 24-27°C. The 24-h incubation period was chosen because there was usually no visible pattern at 4 h and 24 h is regarded as the upper limit of the "active" functional profile, as opposed to the "potential" physiological profile obtained with longer incubations (Garland and Mills, 1991). At the end of incubation, assay plates were placed in plastic bags and refrigerated for up to 3 weeks before being read in a microtiter plate reader. Because the functional profiles determined by this cell extraction procedure were not obtained by following the standard procedure for reliable strain identification (Marello and Bochner, 1989) these profiles should not be compared to those in the BIOLOG® database without risk of misidentification.

Two to eight replicate profiles were obtained and compared to establish the profile at 24 h for each strain in pure culture and also for each sample from the mixed culture experiments. All analyses were done with binary response data (+ or -) obtained by subtracting 140% of the control well's absorbance (Marello and Bochner, 1989) from each well. With that adjustment, positive absorbance values were considered indicative of substrate metabolism, and negative values were considered to indicate no metabolism. Binary patterns were of primary concern in this study because the presence or absence of detectable levels of metabolic activity is the first step in relating population dynamics to the dynamics of metabolic functions. The degree of color development in sample wells is most useful in longer duration studies (Garland and Mills, 1991; Haack et al., 1995). Interval-level absorbance data (i.e., the degree of color development in each of the wells) were used in this study to determine the mean absorbance of substrates giving equivocal responses (for substrate-sample combinations that did not give uniform, e.g. 5/5, results, or for weak color development).

2.3 Application of the perturbation to the community in the bioreactor

In the first experiment, the initial aeration rate ranged between 500 and 800 ml min^{-1}. The community was stressed by ceasing aeration for one week, starting one hour after the Day 22 samples for density and functional profile were removed. At Day 24 an additional functional assay was performed. Minutes after the Day 29 samples were taken, aeration was restarted at the initial rate. Starting at Day 40, aeration became increasingly intermittent and gradually declined in flow rate due to blockage of the outflow line. The outflow air line had to be repeatedly tweaked to reopen an air passage. During the final week of the experiment, the flow rate during bouts of aeration averaged only 400 to 450 ml min^{-1}.

In the second experiment, the initial aeration rate was 400 ± 200 ml min^{-1}. Minutes after the Day 24 removal and replenishment of medium, the community was stressed by ceasing aeration for 5 days. No functional profile sample was removed at Day 24, but an additional functional assay was performed at Day 26, two days into the perturbation. Minutes after the Day 29 density and functional activity samples were taken, aeration was resumed with an air line connected to a new laboratory compressed air supply. A stable aeration rate of about 400 ml min^{-1} was established and maintained until Day 34, when the aeration rate was adjusted to 950-1150 ml min^{-1} after each day's medium removal and replenishment, and ranged between 650-1200 ml min^{-1} over the course of a day. Aeration remained good in the second run until the final day, when the aeration rate fell to 55 ml min^{-1}, at which time the experiment was terminated.

3. Results

3.1 Test reproducibility

To examine the reproducibility of the BIOLOG® assay with the strains of bacteria used in the study, several replicate plates for each strain were inoculated, incubated for 24 h, and read. The binary profiles produced from the replicates were highly reproducible (Table 1). Strain 19 produced the most tests not in complete agreement, with 6 of the 95 compounds producing mixed results. Of the six deviations, 3 had 4/5 plates positve, and 3 had 3/5 plates positive. The other three strains produced a total of 4 non-unanimous results; two tests (1 from strain 9 and 1 from strain 13) had 3/5 plates positive, a single test from strain 13a had 2/5 positive and 1 test from strain 6 had a single plate positive (1/2). In summary, out of 380 total tests replicated as indicated in Table 1, only 10 (2.6%) produced equivocal results. The non-unanimous test results were not clustered within any group of compounds, but were spread throughout the 95 substrates.

Table 1. Tests from BIOLOG®-GN plates in which disagreement among replicates was observed. The remaining 85 of the 95 tests were all in agreement at the level of replication noted

Compound	Strain (No. of replicates)			
	6A (2)	9 (5)	13A (5)	19 (5)
α-cyclodextrin				3
dextrin				4
maltose				4
xylitol				4
α-hydroxybutyric acid				3
p-hydroxyphenylacetic acid		3		
L-proline			3	
L-serine			2	
L-threonine				3
thymidine	1			

In the mixed culture incubations, 9 sampling times produced 855 total tests, each of which was replicated 5 times. In the first incubation, 39/855 tests (4.6%) were not in complete agreement (Table 2). The second incubation had only 16/855 tests (1.9%) that did not agree.

Table 2. Test reproducibility in the mixed communities from the bioreactor incubations. Note that only the tests which were not in complete agreement are shown. All other tests were positive in all replicates at each sampling time or negative in all replicates at each sampling time

Compound	Days in Mixed Culture								
	1	8	15	22	24	29	36	43	50
Tween 80							4		
N-acetyl-D-galactosamine					1				
L-arabinose			4						
L-fucose				1	3			3	
lactulose						4			
D-mannitol				4					
L-rhamnose			4				2		2
mono-methyl succinate		4							
acetic acid		4							
D-gluconic acid			3					3	
D-glucuronic acid	1								
α-hydroxybutyric acid			2						
p-hydroxyphenylacetic acid									3
D,L-lactic acid									2
propionic acid			1		1			2	
succinic acid					1				
L-alanyl-glycine		1	4						
L-proline			2		3				1
L-pyroglutamic acid					1				
L-serine		3							1
L-threonine		2			3			2	
putrescine					2				1
bromosuccinic acid				3					
urocanic acid			1					1	

3.2 Population dynamics and community-level physiological profiles

During the bioreactor incubations, the numbers of cells of the various strains reached an approximate steady state level that was slightly different for each strain (Fig. 1). Strain 19 did not reach the levels of abundance of the other strains except at the very beginning of the experiment. Strain 6A, on the other hand, grew to levels that approximated those of strains 13A and 9, but did so only after about a week of incubation. During the perturbation, the numbers of cells of all of the strains seemed to drop slightly, but those of strain 19 fell by more than an order of magnitude. The die-off of strain 19 in this reactor lagged behind the cessation of aeration and the organism recovered as soon as aeration was reapplied to the reactor.

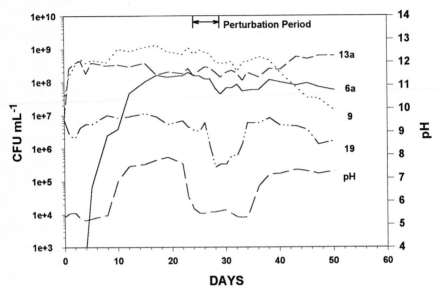

Fig. 1. Dynamics of the inoculated strains in the first bioreactor run. Note that the decrease in countable numbers of strain 19 lagged behind the onset of the perturbation of anoxia

The substrate utilization patterns showed that several of the compounds stopped being metabolized during the perturbation, and that activity did not reappear in all cases after the air was turned on again. Not all were recovered after the reapplication of air to the reactor. For example, metabolism of acetic acid, D-gluconic acid, α-hydroxybutyric acid, p-hydroxyphenylacetic acid, several other carboxylic acids, glycerol, putrescine, and several amino acids ceased during the perturbation, but recommenced after the anoxic period. In still other cases, mostly in the carbohydrates, polymers and esters, activity was present throughout the incubation, including the period of perturbation.

Metabolism of N-acetyl-D-galactosamine, succinic acid, L-alanyl-glycine, L-pyroglutamic acid, and bromosuccinic acid was observed before the perturbation in the bioreactor incubation, but no activity was present after reapplication of the air supply. Because the original isolates were detectable at all times in the reactor, we infer that the genetic makeup of the community did not change so drastically over the course of the experiment to account for the change in the spectrum of compounds metabolized.

3.3. Comparison of community-level physiological profiles with isolate profiles

The number of sole-carbon substrates metabolized by microbial communities changed over time. In the first incubation, the number of degraded carbon sources increased monotonically until the time of the perturbation, and showed a second increase between the end of the perturbation on day 29 and the end of the reactor run. (Fig.2) Due to the slight differences in the binary patterns on the replicate assay plates, the average of the sum of positive responses for each plate peaked at only 44 on day 22, although the total number of active functions was 47 on day 22. According to both measures, the number of active functions dropped to 21 by the end of the perturbation. After the return of aeration, the number of detectable metabolic functions gradually increased to 38. D-glucuronic acid was not degraded in pure culture by any of the isolates, was metabolized at a barely detectable level on the first day, and utilization was never detected in this community again. Five other compounds were also degraded by synergistic interaction among the strains: N-acetyl-D-galactosamine, L-fucose, L-rhamnose, D-gluconic acid, and L-alanylglycine.

In the second run of the mixed community (data not shown), the same 6

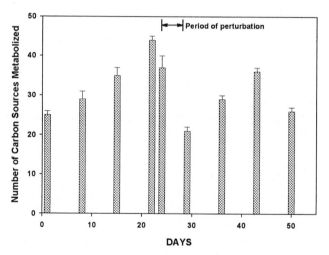

Fig. 2. Changes in the number of positive tests during the course of the first community run in the bioreactor

compounds not degraded by any of the isolates in pure culture were metabolized synergistically by the community, and two additional compounds hydroxy-L-proline and glucose-1-phosphate also were metabolized by the community, although glucose-1-phosphate use was seen only on the first day of the incubation and only in 2 of 5 plates. Of the 30 metabolic activities detected on the first day of the second reactor run, 8 disappeared permanently after the first week. Seven of the 8 affected compounds were synergistically degraded: N-acetyl-D-galactosamine, L-fucose, L-rhamnose, D-glucuronic acid, L-alanyl-glycine, hydroxy-L-proline, and glucose-l-phosphate. Observing the behavior of the community in the first reactor run with respect to those 8 compounds did not clarify possible trends: 2 of the 8 functions never appeared (synergistic metabolism of glucose- 1 phosphate and hydroxy-L-proline), three functions were permanently lost after the perturbation (degradation of N-acetyl-D-galactosamine, L-alanyl-glycine, and D-glucuronic acid), and three were lost and eventually recovered (L-fucose, L-rhamnose, and L-serine).

4. Discussion

The physiological profiling approach to community structure offers a reliable means of examining communities with regard to the arrangement of functional capability of the assemblage. Tests with the BIOLOG® plates were perfectly consistent within replicates in 95% - 99% of all the tests run (2080 for the entire study). While some variability is encountered in dealing with environmental samples (Garland and Mills, 1991; Garland and Mills, 1994), that variability may realistically describe the small-scale patchiness in the distribution of microbes and their functions. Indeed, the variability in environmental samples may prove to be helpful in determining the realistic scales of variability in distribution of microbes that constitute communities.

Use of short-term (i.e., less than 24 hour) incubations of the test plates provides a profile that is more reflective of *in situ* activity, or at least of constitutive activity, possessed by a high proportion of the organisms present in the test well, at least when binary data are considered, because bacteria that utilize wide ranges of carbon and nitrogen sources generally synthesize the appropriate enzymes when they are needed (Mahan *et al.*, 1993). Concerns over the lack of reproducibility of profiles can be overcome by use of binary coding of the data (which minimizes quantitative differences in profiles) combined with replication of the profiles. In natural communities which contain much larger numbers of different organisms than employed in these experiments, one might expect a greater variability of response, but the use of binary data coupled with the threshold cell concentration ($\approx 10^7$ cells ml^{-1}) to effect a visible response in the test well, suggests that variability will be minimal if the sampling scale does not exceed the correlation length scale of the community, i.e., replicate samples are collected from within a single community.

As pointed out by Bouma and Mills (submitted), the signature of a community subjected to a physiological profiling with sole carbon source utilization is not the sum of the characteristic patterns of the individual community members.

Physiological profiling with sole carbon source utilization can be used to demonstrate time-dependent properties of communities. In the present study, several of the substrates metabolized by the communities in the early part of the incubation disappeared during the anoxic period. In some cases those substrates were again metabolized after the resumption of aeration, but others were not. The observation that some of the metabolic capacity of the community was lost permanently is of importance, because none of the organisms inoculated into the reactors became extinct. This behavior is somewhat surprising, as the substrate-nutrient mixture was the same throughout the entire incubation. We cannot determine what caused the permanent loss of the characters.

The ability to detect synergistic relationships among the community members is an important property of the physiological profiling approach to community structure that is not shared by any of the other techniques that are currently applied, including the most sophisticated molecular methods. Neither genetic analysis nor fatty acid profiles can categorize the community on the basis of interactions, either those currently occurring *in situ*, or those potentially present as a result of the activities of the organisms. Perhaps the most powerful community descriptions are those that employ genetic profiling using selected molecular techniques in combination with physiological profiling using approaches such as described here.

5. References

Bouma JE (1995) Dynamics of strain populations and metabolic functions in gnotobiotic microbial communities. Ph.D., University of Virginia, Charlottesville, VA

Bouma JE, Garland JL, Mills AL (submitted) Relationship between population dynamics of bacterial strains and functional dynamics of constructed microbial communities.

Garland JL (1996) Analytical approaches to the characterization of samples of microbial communities using patterns of potential C source utilization. Soil Biol Biochem 28:213-221

Garland JL, Mills AL (1991) Classification and characterization of heterotrophic microbial communities on the basis of patterns of community-level-sole-carbon-source utilization. Appl Environ Microbiol 57:2351-2359

Garland JL, Mills AL (1994) A community-level physiological approach for studying microbial communities. In: Ritz K., Dighton J, Giller KE, (eds) Beyond the Biomass, Wiley, London pp 77-83

Haack SK, Garchow H, Klug MJ, Forney LJ (1995) Analysis of factors affecting the accuracy, reproducibility, and interpretation of microbial community carbon source utilization patterns. Appl Environ Microbiol 61:1458-1468

Insam H, Amor K, Renner M, Crepaz C (1996) Changes in functional abilities of the microbial community during composting of manures. Microb Ecol 31:77-87

Mahan MJ, Slauch JM, Mekalanos JJ (1993) Selection of bacterial virulence genes that are specifically induced in host tissues. Science 259:686-688

Marello TA, Bochner BR (1989) BIOLOG® reference manual: metabolic reactions of gram-negative species BIOLOG® and Science Tech Publishers, Hayward, CA

194

Turco RF, Kennedy AC, Jawson MD (1994) Microbial indicators of soil quality. In: Doran JW, Coleman DC, Bezdicek DF, Stewart BA, (eds) Defining soil quality for a sustainable environment, Soil Science Society of America, Madison, Wisconsin pp 73-90

Winding A (1994) Fingerprinting bacterial soil communities using Biolog microtitre plates. In: K. Ritz, Dighton J, Giller KE, (eds) Beyond the Biomass, Wiley, London pp 84-94

Zak JC, Willig MR, Moorhead DL, Wildman HG (1994) Functional diversity of microbial communities: a quantitative approach. Soil Biol Biochem 26:1101-1108

Biolog Substrate Utilisation Assay for Metabolic Fingerprints of Soil Bacteria: Incubation Effects

Anne Winding and Niels Bohse Hendriksen

Department of Marine Ecology and Microbiology, National Environmental Research Institute, P.O.Box 358, Frederiksborgvej 399, DK-4000 Roskilde, Denmark.

Abstract. Bacterial communities can be described by their enzymatic potentials using the Biolog substrate utilisation assay. We have investigated tetrazolium reduction and cell growth during incubation of *Pseudomonas fluorescens* MM6 and soil bacteria in Biolog plates. Increasing the inoculum size shortened the lag phase before formazan formation. For the soil bacteria, increasing the inoculum also resulted in a higher rate constant of formazan formation, and the final number of wells with formazan formation increased. Both MM6 and soil bacteria proliferated in the wells both with and without specific carbon sources after inoculation. With soil bacteria, the presence of clay, humic substances, and dissolved organic matter increased the background coloration and may have resulted in cell growth. The growth led to increased culturability (CFU/AODC) and rate of colony-appearance and decreased the diversity of the bacterial communities within each well. Metabolic fingerprinting of bacterial communities using Biolog plates thus depends on aerobic growth of a fraction of the community.

Keywords. Biolog substrate utilisation, metabolic fingerprinting, soil bacteria

1. Introduction

To assess the impact of, e.g., use of xenobiotic compounds or released bacteria on the soil environment, it is important to be able to monitor changes in the diversity and activity of bacterial communities. Furthermore, in ecological studies, easy and reliable techniques to describe the structure and dynamics of bacterial communities are needed.

The Biolog substrate utilisation assay for identification of bacteria (Biolog, Inc., Hayward, CA) has been used to characterise intact microbial communities. For instance, it has been possible to differentiate between microbial communities from freshwater, rhizosphere, and nutrient-enriched soil slurries (Garland and Mills 1991, 1994) from agricultural and beech forest soils (Winding 1994), from plant communities (Zak *et al.* 1994), and microbial communities extracted from different sized soil-aggregates (Winding 1994), although Haack *et al.* (1995) were unable to detect differences between rhizosphere and bulk soil microbial communities. Furthermore, the assay has been able to monitor effects of different biotic treatments of soil (Winding *et al.* 1997). Biolog plates have thus been proven useful as a metabolic fingerprint of bacterial communities. However, to interpret the Biolog fingerprints it is essential to know how different inoculum sizes and incubation conditions influence the tetrazolium reduction. In model community studies, the individual strains contributed to the fingerprint by oxidation of specific signature compounds, though the degree of substrate oxidation was not a summation of the individual members abilities (Haack *et al.* 1995, Mills and Bouma 1996). Inoculation of bacteria in the Biolog plates is expected to result in growth, and the rate of formazan development in individual wells has been assumed to represent growth of the inoculated bacteria and to correlate positively with inoculum cell density (Garland and Mills 1991, Winding 1994).

The aim of the present work was to study the effect of incubation conditions on the metabolic activity and growth in the Biolog wells and to evaluate whether specific bacteria were selected for during the Biolog assay.

2. Materials and methods

2.1 Incubation conditions

A *P. fluorescens* strain MM6 (biovar II), isolated from the rhizosphere of barley roots (Michaelsen 1993), was cultured to stationary phase in King's B broth with 10 ml glycerol l^{-1} (Atlas 1993) with agitation (200 rpm) at 15°C for 20-40 h. The cell culture was washed 3 times, resuspended in sterile 0.85% NaCl, and agitated 3-4 hours before inoculation of GN-Biolog plates (Biolog, Inc., Hayward, CA). Soil bacteria were extracted from an agricultural (sandy loam) or a meadow (loam) soil by blending of a 1:10 dilution in Winogradsky salt solution (Holm and Jensen 1977) followed by centrifugation (950 g, 10 min) as described by Winding (1994). The bacterial suspension was inoculated onto GN- or GP-Biolog plates (designed for identification of Gram negative and Gram positive bacteria, respectively). Formazan formation was measured using a video-image analysis system measuring grey level values (Winding 1994). All incubations of Biolog plates were performed at 15°C with agitation at 200 rpm to ensure uniform distribution of the formazan. The total number of MM6 was counted on fixed samples after staining with DAPI (4,6-diamidino-2-phenylindole, Sigma) (Porter and Feig 1980) or by Acridine Orange Direct Count (AODC) (Fluka AG) in the case of soil bacteria (Hobbie *et al.* 1977).

2.2 Effect of incubation time and inoculum density

The effect of incubation time and inoculum size was studied by inoculating 8 different densities of MM6 (from 3.5×10^5 to 8.5×10^7 cells ml^{-1}) and ten-fold dilutions (4 cells ml^{-1} to 4.1×10^6 cells ml^{-1}) of bacteria extracted from the agricultural soil onto GN-Biolog plates. The total cell count and formazan formation were described by the modified Gompertz equation: $y = A*\exp(-\exp((\mu*e)/A*(8-t)+1))$, where $y = \ln(N/N_0)$ where N is cell number or formazan formation at time t, and N_0 the same at time 0, μ = rate constant of growth or formazan formation, $A = \ln(N_4/N_0)$ where N_4 is maximum cell number or formazan formation, and 8 = lag phase (Zwietering et al. 1990), using Sigma Plot ver. 1.02 (Jandel Scientific). Significant correlations were found with Pearson Product Moment Correlation or t-tests using Sigma Stat ver. 1.0 (Jandel Scientific).

During aerobic incubation, extensive growth of the bacteria may cause anaerobic conditions or reduced oxygen availability in the bottom of the wells. To evaluate the effect of reduced oxygen availability, we incubated additional Biolog plates in jars with an anaerobic atmosphere (Anaerocult A, Merck). The inoculum concentrations were 8.5×10^7 cells ml^{-1} and 1.4×10^7 cells ml^{-1} for MM6 and soil bacteria, respectively. In order to increase the number of substrates, the soil bacteria were tested in both GN- and GP-Biolog plates.

The increase in cell number and formazan formation were followed during 186 h of incubation of MM6 (inoculum size: 3.4×10^6 cells ml^{-1}) in 13 wells of a GN-Biolog plate. The wells were chosen to represent different substrate classes and contained carbohydrates (L-arabinose, α-D-glucose, and sucrose), carboxylic acids (D-gluconic acid, D-glucuronic acid, malonate (which MM6 is unable to metabolise), and methyl pyruvate), amino acids (D-alanine, L-histidine, and L-pyroglutamate), Tween 80, or urocanate. The control well without carbon source was also included. Soil bacteria were inoculated into 25 wells of GN-Biolog plates (inoculum size: 2.0×10^7 cells ml^{-1}). The 25 wells represented different substrate classes and contained either carbohydrates (glycogen, L-arabinose, cellobiose, L-fucose, maltose, D-melibiose, D-raffinose, sucrose, glucose-1-phosphate, and glycerol), carboxylic acids (methyl pyruvate, citrate, D-gluconate, ß-hydroxybutyrate, itaconate, propionate, and bromo succinate), amino acids (L-asparagine, L-threonine, amino butyrate, and L-phenylalanine), alaninamide, uridine, phenyl ethylamine, or no specific carbon source (control).

2.3 Diversity in Biolog plates

The diversity of the community in the wells after 3 days of incubation in GP and GN-Biolog plates was compared to the initial soil extract. Nine wells were chosen in order to represent different substrate classes. The carbon sources were carbohydrates (α-D-glucose and sucrose), carboxylic acids (methyl pyruvate, acetate, D-galacturonic acid, ß-hydroxy butyrate, and glycyl-L-glutamate), or amino acids (L-alanine and serine). The control well was also included. Inoculum and cell suspensions in the wells were diluted and plated on Tryptic Soy agar plates (TSA, Difco) with the fungicide natamycin (25 µg ml^{-1}, Merck). After 3 and 14 days of incubation at 15°C, the total number of colonies was counted and the colonies were grouped on basis of morphology. Additionally, the total number of bacteria (AODC) was determined. All colonies appearing after plating of the inoculum and 30 of the colonies appearing after plating the bacterial suspensions of each of the 10 wells of the GN-Biolog plate were randomly selected and characterised by KOH-test for Gram reaction, motility, presence of catalase and oxidase, and oxidative and fermentative abilities on Hugh-Leifson media. A first stage identification of the bacteria was performed according to Barrow and Feltham (1993). Based on colony morphology and characterisation of isolates, Shannon's index of diversity was calculated and significant differences between indices were detected by t-tests according to Zar (1984).

3. Results

3.1 Incubation conditions

The formazan formation after inoculation of *P. fluorescens* MM6 at different concentrations in GN-Biolog plates was modelled by the modified Gompertz equation, but no significant correlations between inoculum concentration and the estimated parameters (rate constant, lag phase, and asymptotic value) were found (data not shown). There was, though, a tendency of a shorter lag phase with increasing inoculum density. The modified Gompertz equation was chosen to describe the data as this equation was found statistically sufficient (Zwietering *et al.*, 1990).

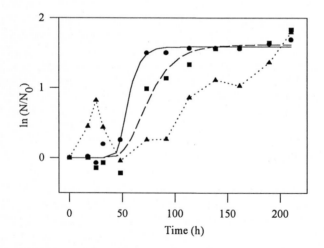

Fig. 1. Normalised sum of formazan formation (ln (N/N$_0$) per Biolog plate as a function of time for different inoculum densities of soil bacteria. For inoculum densities of 4.1 x 10^5 (■) and 4.1 x 10^6 (●) cells ml^{-1} the curve fitting results are shown, while measured data are connected with a dotted line for the inoculum density of 4.1 x 10^4 (▲) cells ml^{-1}.

al. 1990) and superior compared to other equations when describing colour development in Biolog plates (Lee *et al.,* 1995).

An inoculum of >4 x 10^4 cells ml^{-1} of soil bacteria was needed to obtain formazan formation and modelling of formazan formation was only possible with inoculum densities of 4.1 x 10^5 and 4.1 x 10^6 cells ml^{-1} (Fig. 1). The lag phase was significantly shorter and the rate constant of formazan formation significantly higher with higher inoculum density as tested by t-tests ($p<0.05$). Similarly, the total formazan formation in the plates increased and the number of wells with significant formazan formation increased from 12 to 71. Free soil enzymes included in the soil extract may reduce tetrazolium violet (Tabatai 1982). In our study, chloramphenicol, added in a concentration found to inhibit enzyme synthesis and growth (400 µg ml^{-1}), led to absence of formazan formation (data not shown). Therefore, the tetrazolium reduction detected is based on microbial activity and not on soil enzymes present at the time of incubation. Addition of the fungicide natamycin to the inoculum did not alter the colour development, an indication that fungi had no effect on the Biolog assay in this study.

The increase in total number of MM6 and formazan formation as a function of incubation time showed a typical growth pattern for the 12 different carbon sources tested and the control well (exemplified in Fig. 2, left-hand column). In the control well an initial increase in cell abundance to 1 x 10^8 cells ml^{-1} without concomitant formazan formation was found and is assumed to be based on the peptone and yeast extract present in the wells (Bochner 1978). When a carbon source was present, cell growth continued to a maximum of 3 x 10^9 cells ml^{-1}. Simultaneously, formazan formation increased as a function of time (Fig. 2, left-hand column). In the case of malonate, which *P. fluorescens* cannot oxidise, the cell abundance increased to an intermediate level of 5 x 10^8 cells ml^{-1} while very low formazan formation was observed (data not shown). The rate constant of formazan formation showed a significant negative correlation with the growth rate constant ($p<0.05$), while the asymptotic value, A, of formazan formation showed a positive correlation with the A value of growth ($p<0.05$). The lag phase of formazan formation and cell growth did not correlate.

When soil bacteria were inoculated onto GN-Biolog plates, cell numbers increased in all 25 wells from 2.0 x 10^7 to a maximum of 1.6 x 10^9 cells ml^{-1} during the 96 h incubation. The number of cells in the control well reached the same level as in the other wells. Cell growth took place after a lag phase of approximately 36 h followed by formazan formation after a lag phase of approximately 48 h (examples shown in Fig. 2, right-hand column). For 20 of the 25 carbon sources tested, cell growth and formazan formation could be modelled by the modified Gompertz equation, but the estimated parameters showed no significant or meaningful correlations.

At reduced oxygen availability, the formazan formation by MM6 was 43% lower than at aerobic conditions (data not shown). The same wells showed formazan formation except those with L-alanine, L-alanylglycine and N-acetyl-D-glucosamine. The lower formazan formation was thus mainly due to a quantitative reduction of activity. Contrary to this, the soil bacteria did not show detectable formazan formation at reduced oxygen availability in neither GN- nor GP-Biolog plates (data not shown). Prolonged incubation (7 days) of the plates at reduced oxygen conditions resulted in accidental formazan formation in a few wells, indicating slow growth of the bacteria.

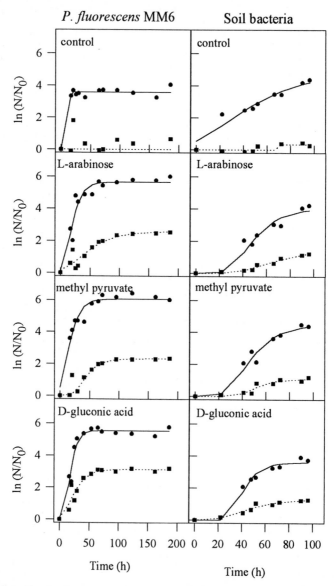

Fig. 2. Normalised cell number and formazan formation ($\ln(N/N_0)$)) during incubation of *P. fluorescens* MM6 (left-hand column) and soil bacteria (right-hand column) on different carbon sources in GN-Biolog plates. ●: cell number; ■: formazan formation.

Table 1. Colony forming units (CFU), total number of bacteria (AODC), and Shannon's diversity index based on morphological different colonies and first-stage identified isolates before (inoculum) and after incubation of soil bacteria in 10 wells of GN- and GP-Biolog plates.

Biolog plate		CFU log cells ml^{-1} (+/-3-8%)[a]	AODC (+/- 7%)[a]	Diversity colonies H'	isolates H'
	inoculum	4.04	6.53	0.912	0.975
GN-Biolog	control	8.00	8.34	0.699*	0.588*
	α-D-glucose	8.73	9.27	0.291*[+]	0.498*
	sucrose	8.70	9.26	0.268*[+]	0.725*
	methyl pyruvate	8.67	9.13	0.123*[+]	0.698*
	acetate	8.35	8.93	0.466*[+]	0.576*
	D-galactonate	8.49	9.26	0.300*[+]	0.414*
	ß-hydroxy butyrate	8.50	8.62	0.255*[+]	0.355*[+]
	L-alanine	8.59	9.16	0.394*[+]	0.203*[+]
	glycyl-L-glutamate	8.60	8.71	0.446*[+]	0.497*
	D-serine	7.54	7.65	0.178*[+]	0.342*[+]
	all isolates	na	na	na	0.694*
GP-Biolog	control	7.67	8.08	0.546*	nd
	α-D-glucose	8.51	8.89	0.357*[+]	nd
	sucrose	8.12	8.88	0.402*[+]	nd
	methyl pyruvate	8.65	9.06	0.514*	nd
	acetate	8.25	8.54	0.433*[+]	nd
	D-galactonate	8.22	8.89	0.234*[+]	nd
	ß-hydroxy butyrate	7.94	8.45	0.403*[+]	nd
	L-alanine	8.34	8.99	0.474*[+]	nd
	glycyl-L-glutamate	8.18	8.71	0.450*[+]	nd
	L-serine	8.44	9.27	0.284*[+]	nd

na: not available
nd: not determined
[a]:standard deviations according to the Poisson distribution ($\sqrt{n}/n$).
*: significantly different from inoculum (p<0.05).
[+]: significantly different from control (p<0.05).

3.2 Diversity in Biolog plates

The total and the culturable number of bacteria increased by a factor of 10^3 and 10^4, respectively, after incubation in Biolog plates (Table 1) and the culturability (CFU/AODC) increased from 0.3% in the inoculum to 15-78% in the wells after incubation. The agar plates with the inoculum had to incubate for 14 days to reach the maximum number of CFUs, while after incubation in the Biolog plates the agar plates only had to incubate for 3 days. The rate of colony appearance thus increased after incubation in the Biolog plates. Shannon's index of diversity based on colony morphology and first-stage identification of isolates showed significantly higher diversity of the inoculum than of the cell suspensions in the wells (Table 1). The diversity of all the isolates from the GN-plate pooled together was also significantly lower than the diversity of the inoculum. The diversity based on colony morphology was significantly higher for the control well than for the wells with specific carbon sources, except the well with methyl pyruvate in the GP-Biolog plate. The species composition based on first-stage identification of the isolates was significantly different among the inoculum and the isolates from the plate as tested by a chi-square test. Among the inoculum isolates no single genera dominated, while among the other isolates more than 50% were assigned to the genera of *Pseudomonas, Achromobacter, Agrobacterium* or *Janthinobacterium* and almost 15% were assigned to the genera of *Chromobacterium, Vibrio, Plesiomonas,* or *Aeromonas*.

4. Discussion

4.1 Incubation conditions

The carbon sources in the wells have been chosen to distinguish bacterial species and therefore, each carbon source is expected to be oxidised by a limited number of species in a soil population. By increasing the inoculum size, an increasing number of carbon sources was oxidised leading to a shorter lag phase and a higher rate constant of formazan formation (Fig. 1) as also found by Garland (1996) and Haack *et al.* (1995). In our hands, the lag phase could not be reduced further by inoculating more cells as this resulted in coloration of the control well, due either to coloured humic substances or oxidation of dissolved soil organic matter and concomitant formazan formation. This was also found by Garland and Mills (1994) who never inoculated more than 10^5 to 10^7 cells ml^{-1}. Furthermore, they (Garland and Mills 1991) compensated for variation in the tetrazolium reduction by dividing the grey level values of each well with the average well colour development.

Growth took place in all wells prior to formazan formation for both MM6 and soil bacteria (Fig. 2). When the pure culture was able to oxidise the carbon source, growth continued to a higher level than when no or a non-utilisable carbon source was included. The growth of both MM6 and soil bacteria increased to approximately the same level of 1×10^9 cells ml^{-1} when the carbon source could be oxidised. The soil bacteria always increased in numbers to this level, even in the control well and in wells that did not result in formazan formation. This growth must be based on peptone and yeast extract, added to the wells to support growth without formazan formation (Bochner 1978), in addition to the aforementioned factors of increasing coloration of the wells. When the carbon source could be oxidised, a cell growth up to approximately 10^8 CFU ml^{-1} was reported (Haack *et al.* 1995). This ten times lower cell abundance might be due to low culturability, as we found culturabilities of 15-78% (Table 1). Growth was also reported by Garland and Mills (1991), whereas Ellis *et al.* (1995) reported little growth to occur in the wells and found that less than 10 CFUs were required for a positive reaction. With a culturability of 0.3% (Table 1), which is the lowest culturability we have observed in this soil, at least 18 CFUs were added. Growth of the inoculated bacteria to $>10^8$ cells ml^{-1} prior to formazan formation thus seemed to be a prerequisite for fingerprinting.

For the pure culture, the asymptotic values, A, for growth and formazan formation correlated ($p<0.05$), respectively. Such a correlation was not found by Haack *et al.* (1995). This discrepancy may be due to different bacterial strains as the extent of tetrazolium reduction is dependent on many physiological and chemical parameters (Thom *et al.* 1993). The lack of correlations between estimated parameters for the soil bacteria is probably caused by the initial growth in all wells including the control well, a growth that seems uncoupled to formazan formation. In support of this, it has been shown that not all bacteria are able to reduce tetrazolium salts (Winding *et al.* 1994) and that no simple relationship between soil respiration and dehydrogenase activity measured by tetrazolium reduction exists (Howard 1972).

MM6 is able to denitrify, and the reduced tetrazolium reduction at reduced oxygen availability may be supported by an active electron transport system during reduced oxygen availability. The growth yield of a *P. denitrificans* at denitrifying conditions was found to be approximately half of that obtained at aerobic conditions (Koike and Hattori 1975). At reduced oxygen availability, soil bacteria were apparently not able to reduce detectable amounts of tetrazolium, unable to grow to a density enabling detectable formazan formation or both. If aerobic respiration during the incubation in the wells causes limiting oxygen conditions, our results indicate an arrest in formazan formation. Bacteria able to reduce tetrazolium violet at reduced oxygen availability are thus of minor influence, when soil bacteria are tested in Biolog plates.

4.2 Diversity in Biolog plates

The number of different colonies and isolates were never reduced to one in a single well as would be expected in an enrichment culture. Furthermore, the dominating colonies and isolates were different on the different carbon sources (data not shown). This can be due to the duration of the experiment or the carbon sources in the wells which might be oxidised by several bacterial species and give room for development of bacterial consortia with a broader functional diversity than single strains (Zak *et al.*, 1994). The rate of colony formation and the culturability increased after incubation in Biolog plates, though in some wells the culturability was low (20%). As the cell number had increased by a factor of 10^4, the non-culturable bacteria counted after incubation in the wells must have proliferated in the wells. Therefore, it is probable that the viability in the wells has been higher than the culturability, which again might be due to the selected agar media. In our experiments, the non-culturable as well as the culturable bacteria have thus contributed to the formazan formation. The first-stage identification of the isolates revealed a change towards species often isolated on agar plates and towards bacteria which can grow on many different substrates at a wide range of concentrations. The culturable bacteria reducing tetrazolium salt in the Biolog plate are thus generally fast-growing bacteria growing at high nutrient concentrations. Slow-growing bacteria unable to grow at high nutrient concentrations may be neglected in the assay. In combination with the observation that the degree of substrate oxidation by a community is not a summation of the oxidation by individual members (Haack *et al.* 1995, Mills and Bouma 1996), it can be concluded that the metabolic profile of a microbial community obtained by the Biolog substrate utilisation assay is the result of the growth and metabolic activity of only a fraction of the microbial community.

5. Acknowledgement

Technical assistance was provided by BR Hansen and JK Skalshøi. Dr. O Nybroe is acknowledged for donation of specific antibodies against DF57. Dr. N Kroer is acknowledged for critical comments on the manuscript. This research was supported by the Danish Center for Microbial Ecology and the Danish Environmental Protection Agency (file no. M 6045-0016).

6. References

Atlas RM (1993) Handbook of Microbiological Media (LC Parks) p 472. CRC Press, Inc. Boca Raton, Florida

Barrow GI, Feltham RKA (1993) Cowan and Steel's Manual for the identification of medical bacteria. 3. ed. Cambridge University Press

Bochner BR (1978) Device, composition and method for identifying microorganisms. United States Patent no. 4,129,483

Ellis RJ, Thompson IP, Bailey MJ (1995) Metabolic profiling as a means of characterizing plant-associated microbial communities. FEMS Microbiol Ecol 16: 9-18

Garland JL, Mills AL (1991) Classification and characterization of heterotrophic microbial communities on the basis of patterns of community-level sole-carbon-source utilization. Appl Environ Microbiol 57: 2351-2359

Garland JL, Mills AL (1994) A community-level physiological approach for studying microbial communities. In: Ritz K, Dighton J, Giller K (eds.) Beyond the Biomass, Wiley and Sons, Chichester, pp. 77-83

Garland JL (1996) Analytical approaches to the characterization of samples of microbial communities using patterns of potential C source utilization. Soil Biol Biochem 28:213-221

Haack SK, Garchow H, Klug MJ, Forney LF (1995) Analysis of factors affecting the accuracy, reproducibility, and interpretation of microbial community carbon source utilization patterns. Appl Environ Microbiol 61:1458-1468

Hobbie JE, Daley RJ, Jasper S (1977) Use of Nuclepore filters for counting bacteria by fluorescence microscopy. Appl Environ Microbiol 33: 1225-1228

Holm E, Jensen V (1972) Aerobic chemoorganotrophic bacteria of a Danish beech forest. Oikos 23:248-260

Howard PJA (1972) Problems in the estimation of biological activity in soil. Oikos 23: 235-240

Koike I, Hattori A (1975) Growth yield of a denitrifying bacterium, *Pseudomonas denitrificans*, under aerobic and denitrifying conditions. J Gen Microbiol 88: 1-10

Lee C, Russell NJ, White GF (1995) Rapid screening for bacterial phenotypes capable of biodegrading anionic surfactants: development and validation of a microtitre plate method. Microbiology 141:2801-2810

Michaelsen MN (1993) Bacterial populations in the rhizosphere of barley. The Royal Veterinary and Agricultural University. M.Sc. Thesis. (in Danish)

Mills AL, Bouma JE (1996) Strain and functional stability in gnotobiotic reactors. Abstracts of the SUBMECO Conference, Innsbruck, Austria, Oct. 16-18, 1996

Porter KG, Feig YS (1980) The use of DAPI for identifying and counting aquatic microflora. Limnol Oceanogr 25: 943-948

Tabatabai MA (1982) Soil enzymes. In: Page AL, Miller RH, Keeney DR (eds.) Methods of soil analysis, part 2 - Chemical and microbiological properties, Madison, Wisconsin, pp. 903-943

Thom SM, Horobin RW, Seidler E, Barer, MR (1993) Factors affecting the selection and use of tetrazolium salts as cytochemical indicators of microbial viability and activity. J Appl Bact 74: 433-443

Winding A (1994) Fingerprinting bacterial soil communities using Biolog microtitre plates. In: Ritz K, Dighton J, Giller K (eds.) Beyond the BiomassWiley and Sons, Chichester, pp. 85-94

Winding A, Binnerup SJ, Sørensen J (1994) Viability of indigenous soil bacteria assayed by respiratory activity and growth. Appl Environ Microbiol 60: 2869-2875

Winding A, Rønn R, Hendriksen NB (1997) Bacteria and protozoa in soil microhabitats as affected by earthworms. Biol Fertil Soils (in press)

Zak JC, Willig MR, Moorhead DL, Wildman HG (1994) Functional diversity of microbial communities: a quantitative approach. Soil Biol Biochem 26: 1101-1108

Zar JH (1984) Biostatistical analysis. Prentice Hall, New Jersey

Zwietering MH, Jongenburger I, Rombouts FM, van't Riet K (1990) Modeling of the bacterial growth curve. Appl Environ Microbiol 56: 1875-1881

A Novel Approach for Assessing the Pattern of Catabolic Potential of Soil Microbial Communities

Bradley Degens

Department of Environmental Sciences, University of East London, Romford Road, London E15 4LZ, Great Britain

Keywords. Metabolic diversity, substrate induced respiration, substrate catabolism, method development

1. Introduction

Little is known of the importance of the diversity of soil microbial communities in the functioning of soil systems (Meyer 1993; Beare *et al.* 1995; Moore *et al.* 1995). In most investigations of soil microbial communities, diversity is only characterized to the point of distinguishing fungi and bacteria populations (Tunlid and White 1992; Swift and Anderson 1993) or comparisons of whole soil DNA hetrogeneity (Torsvik *et al.* 1990; Ritz and Griffiths 1994). There are currently no techniques that can reliably assess the functional diversity of the whole soil microbial community in soils.

A technique using Biolog microtiter plates has been widely reported to assess patterns of functional potential of soil communities using patterns of substrate utilization (Garland and Mills 1991; Zak *et al.* 1994). The catabolic potential of soil microbial communities is assessed by determining the growth of suspensions of soil organisms in Biolog microtiter plates loaded with 95 simple carbon substrates (Garland and Mills 1991; Zak *et al.* 1994; Haack *et al.* 1995). Microbial species composition and numbers can change during growth in the microtiter wells (Winding 1994) which raises doubts about the reliability of the technique in measuring metabolic properties of *in situ* microbial populations. An alternative approach may be to directly measure the capacity of organisms to use simple organic compounds by measuring substrate induced respiration (SIR). The SIR of micro-organisms over 0-6 hours is characteristic of the lag-phase microbial population in soils (Anderson and Domsch 1973; Anderson and Domsch 1978). Measurement of the SIR of soil micro-organisms with a broad range of substrates may provide a rapid measure of the pattern of catabolic potential of microbial communities in soils.

This paper presents a novel approach to measuring patterns of catabolic potential of soil microbial communities based on the SIR technique. The effect of substrate concentrations on the SIR of microbial communities was investigated in 5 soils over 4 hours. In subsequent experiments, the SIR of soils to 83 substrates (at concentrations established in the first stage) were determined to identify those substrates giving the greatest differences between soils. These substrates were used to assess the changes in catabolic diversity of microbial communities in a soil following addition of glucose.

2. Materials and Methods

Bulk samples of moist topsoil (0-10cm) of each soil were collected from sites under long-term grass pasture (IGER pasture, ADAS pasture, lake pasture), arable cropping (ADAS arable) and spruce forest (lake forest). The soils ranged in organic content from 0.7% in the ADAS arable to 4.4% in the lake grass pasture and all contained more than 30% clay with pH between 3.8 in the lake forest to 6.6 in the ADAS arable.

In preliminary tests, substrates from each of the main classes of chemicals (Table 1) were tested on four soils to determine appropriate concentrations of substrates suitable for testing the SIR of soils to a wider range of substrates. The list of chemicals is a combination of those used in Biolog microtiter plates (Zak *et al.* 1994) and compounds found in the organic fraction of soils. Each substrate (including water control) was added as 2 ml solutions to 1 g equivalent dry weight of soil in McCartney bottles, sealed and incubated for 4 hours at 25°C. The bottles were mixed regularly to ensure efficient exchange of CO_2/O_2 between the slurry and headspace gas. The CO_2 content of the headspace gas was determine using gas chromatography. Twelve substrates were investigated in the ADAS pasture, IGER pasture, Lake Pasture and Forest soils at three rates: D-glucose, D-galactose, lactose and D-mannose at 25, 50 and 100 mM; L-alanine and L-glutamic acid at 5, 10 and 20 mM; Na-citrate, α-ketoglutaric acid and malic acid at 75, 150 and 225 mM; inosine at 2, 10 and 20 mM; glycerol at 5, 20 and 100 mM; and Tween 80 at 10, 20 and 40 mM (approximate molarities as the Tween 80 mixture contained 70% oleic acid with the balance made up by linoleic, palmitic and stearic acids). These concentrations were chosen based on the results of preliminary investigations. CO_2 efflux in the head-space of the ADAS and IGER soils was also analysed at three times (1, 2.5 and 5 hours) using the substrate concentrations providing maximum SIR responses (as determined in the above section).

Substrate responses to 83 substrates comprising a range of amino acids, carbohydrates, organic acids, alcohols, amines and amides (Table 1) were determined for five soils using substrate concentrations as determined in the preliminary tests. These analyses were conducted using three replicate samples for each substrate. The 36 substrates giving greatest differences between these soils were chosen and similar tests applied to ADAS arable soil at 7, 17 and 31 days after amendment with 200 μg glucose-C and 25 μg NH_4NO_3-N per gram dry soil. This treatment was compared with non-amended soil. All of the assays were conducted over four hours and testing of 36 substrates for each soil was

Table 1. The 83 organic substrates screened for differences in SIR responsiveness between soils.

	Organic substrates	
Amino acids	**Amides**	**Carboxylic acids**
χ-aminobutyric acid	L-alaninamide	acetic acid
L-alanine	glucuronamide	ascorbic acid
DL-alanylglycine	succinamide	Na-citrate
L-apsaragine		Na-formate
L-arginine	**Amines**	fumaric acid
L-aspartic acid	D-glucosamine	D-galactonic acid
L-cysteine	L-glutamine	D-gluconic acid
D-glutamic acid	N-methyl-D-glucamine	D-glucuronic acid
L-glutamic acid	N-acetyl-D-glucosamine	DL-α-hydroxybutyric acid
L-glycine	putrescine	DL-β-hydroxybutyric acid
glycyl-L-phenyalanine		α-ketobutyric acid
DL-histidine	**Aromatic chemicals**	α-ketoglutaric acid
hydroxy-L-proline	inosine	α-ketovaleric acid
L-leucine	urocanic acid	DL-lactic acid
L-lysine		maleic acid
DL-methionine	**Carbohydrates**	malic acid
L-ornithine	D-arabinose	malonic acid
L-phenylalanine	L-arabitol	methylsuccinic acid
L-proline	i-erthritol	oxalic acid
DL-pyroglutamic acid	D-fructose	pantothenic acid
L-serine	D-galactose	propionic acid
DL-threonine	D-glucose	quinic acid
DL-tryptophan	inositol (meso)	succininc acid
DL-tyrosine	DL-lactose	tartaric acid
	D-mannose	uric acid
	melibiose	valeric acid
Alcohols	raffinose	
2,3-butanediol	L-rhamnose	**Polymers**
glycerol	sucrose	α-cyclodextrin
mannitol	D-trehalose	dextran (MW 12 000)
D-sorbitol	D-xylose	Tween 80
xylitol		

determined over two consecutive days. All microbial biomass C measurements were determined by the fumigation extraction technique (Sparling *et al.* 1993). Principal components analyses were determined on standardised data using SPSS for Windows (Version 6.0.1).

3. Results

3.1 Substrate Concentrations and SIR Rates Over Time

Maximum respiration rates were generally achieved over a two-fold concentration range for the majority of substrates in most of the soils. These ranges were: 50-100 mM for the carbohydrates, 10-20 mM for the amino acids and inosine, 140 and 210 mM for the carboxylic acids, 20-100 mM for glycerol and 20-40 mM for Tween 80 (Degens, unpublished data). For each class of chemicals, the mid point of the concentration ranges giving maximum SIR responses were used to determine the SIR responses of each chemical class in subsequent measurements. These concentrations were: 15 mM for amino acids, amines and amides; 60 mM for alcohols; 15 mM for aromatic chemicals; 75 mM for carbohydrate compounds; 190 mM for carboxylic acids and 30 mM for the polymers (Degens, unpublished data).

In general, maximum SIR rates for most substrates were attained after 1 hour of incubation in the ADAS and IGER pasture soils (Table 2; IGER pasture soil not presented). This demonstrated that an assay time of 4 hours was reasonable for assessment of SIR of soils using a wider range of substrates. SIR rates consistently increased over time only with the addition of α-ketoglutaric acid in ADAS pasture soil (Table 2). However, the proportional increases in respiration were not exponential at the end of the incubation period, which would have indicated rapid microbial growth on the added substrate.

3.2 SIR Profiles of Different Soils

There were substantial differences in the SIR responses of the 5 soils to 83 substrates. The differential SIR profiles effectively separated the soils in the plot of the first two principal components (Fig. 1a). However, the pattern of separation was not consistent with land use. The SIR profiles of ADAS arable and lake pasture soils were similar, but separated from the forest, ADAS pasture and IGER pasture soils (Fig. 1a).

The 36 most variable substrates across the five soils were selected for use in determining the SIR profiles in soils analysed in later parts of this investigation. These included: 1 amide (succinamide), 3 amines (glucosamine, glutamine, *N*-methyl-glucamine), 10 amino acids (arginine, asparagine, cysteine, glutamic acid, histidine, leucine, lysine, phenylalanine, serine, tyrosine), 1 aromatic

Table 2. Respiration rates of ADAS pasture (mean± standard error in brackets) over 0-1, 1-2.5 and 2.5-5 hours after amendment with distilled water or organic substrates (n=4).

Substrate	Conc.	Respiration rates (μg CO_2 g^{-1} soil h^{-1})					
		0-1h		1-2.5h		2.5-5h	
Distilled water		21.7	(5.0)	17.0	(1.6)	11.2	(1.5)
D-Galactose	75	37.8	(8.5)	25.2	(7.9)	31.9	(2.5)
D-Glucose	75	40.0	(5.1)	61.0	(2.8)	41.6	(4.8)
DL-Lactose	75	23.5	(1.3)	27.4	(5.3)	31.1	(2.4)
D-Mannose	75	52.5	(4.4)	35.2	(2.9)	33.0	(1.4)
L-Alanine	15	43.7	(4.0)	24.4	(3.5)	24.0	(1.6)
L-Glutamic acid	15	17.1	(1.9)	24.5	(5.7)	21.8	(1.6)
Na-citrate	190	34.1	(2.1)	39.8	(3.2)	34.9	(3.6)
Malic acid	190	32.6	(8.3)	35.4	(1.0)	31.5	(4.8)
α-Ketoglutaric acid	190	204.8	(17.9)	161.6	(79.3)	273.4	(53.2)
Glycerol	60	29.2	(5.7)	14.9	(5.2)	24.2	(0.6)
Inosine	15	36.7	(6.3)	37.8	(3.3)	27.6	(1.7)
Tween 80	30	21.8	(2.3)	82.2	(5.6)	77.0	(21.1)

chemical (urocanic acid), 2 carbohydrates (glucose, mannose), 17 carboxylic acids (ascorbic acid, Na-citrate, fumaric acid, Na-formate, gluconic acid, α-hydroxybutyric acid, α-ketobutyric acid, α-ketoglutaric acid, α-ketovaleric acid, malic acid, malonic acid, oxalic acid, pantothenic acid, quinic acid, succininc acid, tartaric acid, uric acid) and 2 polymers (α-cyclodextrin, tween 80). Only 36 were chosen because this was the maximum number of SIR assays that could be determined over two consecutive days. Those substrates for which the SIR responses were not consistently greater than the respiration measured when only distilled water was incubated in the soils were not included. Lactic acid and β-hydroxybutyric acid were also excluded to allow the inclusion of a broader range of amino acids.

3.3 Short-Term Changes in SIR Profiles

Addition of glucose to ADAS arable soil caused distinct changes in the SIR profiles (as revealed in the principal components analysis of these results) which were apparent at 7 days (Fig. 1b). Over subsequent days, the response profiles of the amended soil converged toward that of the non-amended soil (Fig. 1b). The changes in SIR profiles caused by the amendment treatment were much less than differences between the SIR profiles of the 5 soils (Fig. 1c). Glucose addition also resulted in increases in microbial biomass C from 190 μg C g^{-1} soil at day 0

Fig. 1. Position in relation to the first two principal components of the SIR profiles of (a) 5 soils (using 83 substrate SIR profiles); (b) ADAS arable soil (using 36 substrate SIR profiles) over 31 days after amendment with glucose; and (c) 5 soils in relation to the ADAS arable soil amended with glucose (using 36 substrate SIR profiles).

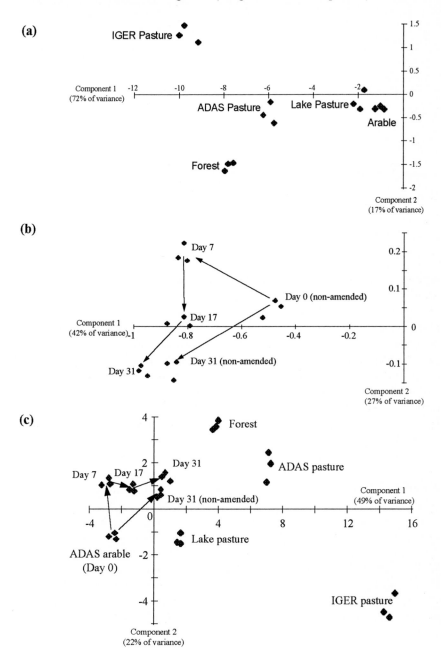

to 257 µg C g^{-1} soil at 7 days (P<0.05), thereafter declining to 217 µg C g^{-1} soil at 31 days. However, the changes in glucose SIR responses and microbial biomass C contents over time were correlated only with the SIR responses to tartaric acid, Na-formate, uric acid, glutamine and α-cyclodextrin (P<0.05).

4. Discussion

The physiological method that has been described provides a unique approach for distinguishing patterns of catabolic potential of microbial communities in soils without prior extraction or culturing of the organisms. This approach is likely to provide a more direct measure of patterns of catabolic potential than culture-based methods such as the Biolog system largely because of the reliance on measuring the metabolic competence of the indigenous microbial population in soils. The technique was sufficiently sensitive to distinguish changes in the pattern of catabolic potential occurring over short time periods as well as large differences in patterns that had developed in field soils over many years. At this stage, the 36 substrates used to detect differences in diversity between a wide range of soils serve as a **guideline** only for future investigations. In addition, the SIR profiles also may be useful in monitoring the recovery of soils to initial states after physical or chemical disturbances. The method can probably be further refined (using different substrates) to identify substrates able to detect fine differences between soils under similar land use.

Most of the SIR responses to the 36 selected substrates were not correlated with changes in microbial biomass C, indicating that the SIR responses identified differences in the characteristics of microbial communities that were not simply due to differences in mass of organisms in soils. SIR responses to glucose can be directly proportional with the total microbial biomass in a wide range of soils (Anderson and Domsch 1978; Sparling and Williams 1986; Sparling and West 1988). The technique described here depends on the differences in SIR responses of those substrates not closely related to the total biomass of microbial communities and probably identifies groups of organisms (the mass of which may not be directly related to the total mass of organisms in soils). This shows that differences between SIR profiles are not strongly weighted by differences in biomass, but reflect qualitative differences between microbial communities. The relative differences in SIR responses to different sugars were not large between the five test soils and were much less than relative differences found between the SIR responses for other organic substrates. Patterns of inducible carbohydrate catabolism in soil micro-organisms are similar probably because of the fundamental importance of this metabolic process during the early stages of competition for organic resources in soils. Consequently, carbohydrate SIR responses correlate more closely with microbial biomass than processes associated with the catabolism of other organic substrates.

Changes in SIR profiles over time may reflect changes in the composition of mineralisable organic matter in soil. The SIR of microbial communities to specific amino acids can be selectively primed by pre-incubation of the specific amino acid in soil (Hopkins *et al.* 1994). Changes in the SIR of microbial communities over time in similar soil types under different management may reflect changes in the composition of the organic C mineralized by microbes. The SIR profile of microbial communities can remain quite stable, since there was little change in the SIR profile of ADAS arable soil 7 days after addition of glucose. This probably reflected the minimal change in the composition of the organic C mineralized by the micro-organisms growing on the glucose-derived organisms. There was little evidence of the glucose-derived organisms at 7 days which is not unexpected since the glucose was probably consumed within 2-3 days and the initial organisms were subsequently attacked by secondary populations of organisms.

Differences between the SIR profiles of soils from different sites must be interpreted with caution and are probably not solely due to differences in the types of organic C mineralized by microbial communities. The differences between the SIR profiles of the ADAS arable soil amended with glucose were smaller than between the different soil types under continuous pasture. It was probable that differences between the physical and chemical properties of the soils had a bearing on the SIR profiles in addition to differences between the forms of mineralisable organic matter in the soils.

5. Acknowledgements

I thank Dr Daniel Murphy of the Institute of Arable Crops Research-Rothamsted, and Dr Anne Bhogal of the Agriculture, Development and Advisory Service, Gleadthorpe, for providing and collecting the soils.

6. References

Anderson JPE, Domsch KH (1973) Quantification of bacterial and fungal contributions to soil respiration. Arch für Mikrobiol 93: 113-127

Anderson JPE, Domsch KH (1978) A physiological method for the quantitative measurement of microbial biomass in soils. Soil Biol Biochem 10: 215-221

Beare MH, Coleman DC, Crossley Jr. DA, Hendrix PF, Odum EP (1995) A hierarchical approach to evaluating the significance of soil biodiversity to biogeochemical cycling. Plant Soil 170: 5-22

Garland JL, Mills AL (1991) Classification and characterization of heterotrophic microbial communities on the basis of patterns of community-level sole-carbon-source utilization. Appl Environ Microbiol 57: 2351-2359

Haack SK, Garchow H, Klug MJ, Forney LJ (1995) Analysis of factors affecting the accuracy, reproducibility and interpretation of microbial community carbon source utilization patterns. Appl Environ Microbiol 61: 1458-1468

Hopkins DW, Isabella BL, Scott SE (1994) Relationship between microbial biomass and substrate induced respiration in soils amended with D- and L-isomers of amino acids. Soil Biol Biochem 26: 1623-1627

Meyer O (1993) Functional groups of micro-organisms. In: Schultz E-D and Mooney HA (eds) Biodiversity and Ecosystem Function, Springer-Verlag, London pp. 67-96

Moore R, Anderson AR, Ray D, Walker C, Pyatt DG, Evans HF, Carter CI, Straw NA, Wainhouse D, Winter TG (1995) Soil biodiversity: a literature review. Scottish Natural Heritage. No 17

Ritz K, Griffiths BS (1994) Potential application of a community hybridization technique for assessing changes in the population structure of soil microbial communities. Soil Biol Biochem 26: 963-971

Sparling GP, West AW (1988) Modifications to the fumigation-extraction technique to permit simultaneous extraction and estimation of soil microbial C and N. Comm Soil Sci Plant Anal 19: 327-334

Sparling GP, Williams BL (1986) Microbial biomass in organic soils: estimation of biomass C and effect of glucose or cellulose amendments on the amounts of N and P released by fumigation. Soil Biol Biochem 18: 507-513

Sparling GP, Gupta VVSR, Zhu CY (1993) Release of ninhydrin-reactive compounds during fumigation of soil to estimate microbial C and microbial N. Soil Biol Biochem 25: 1803-1805

Swift MJ, Anderson JM (1993) Biodiversity and ecosystem function in agricultural systems. In: Schultz E-D and Mooney HA (eds) Biodiversity and Ecosystem Function, Springer-Verlag, London, pp. 15-41

Torsvik V, Salte K, Sorheim R, Goksöyr J (1990) Comparison of phenotypic diversity and DNA heterogeneity in a population of soil bacteria. Appl Environ Microbiol 56: 776-781

Tunlid A, White DC (1992) Biochemical analysis of biomass, community structure and metabolic activity of microbial communities in soil. In: Stotzky G and Bollag JM (eds) Soil Biochemistry, Volume 7. Marcel Dekker, New York, , pp. 229-262

Winding A (1994) Fingerprinting bacterial soil communities using Biolog microtiter plates. In: Ritz K, Dighton J and Giller KE (eds) Beyond the Biomass, Wiley-Sayce, London, pp. 85-94.

Zak JC, Willig MR, Moorhead DL, Wildman HG (1994) Functional diversity of microbial communities: A quantitative approach. Soil Biol Biochem 26: 1101-1108

Chirality is a Factor in Substrate Utilization Assays

David W. Hopkins and Roger W. O'Dowd

Department of Biological Sciences, University of Dundee, Dundee DD1 4HN, UK

Abstract. We have compared the rates of metabolism of the D- and L-enantiomers of several amino acids in soils with a range of biological and chemical properties. The short-term substrate induced respiration (SIR) rates when the D-amino acids were added to soils were less than those of the corresponding L-amino acids. L-alanine, -glutamine and -glutamic acid SIRs were reasonable predictors of soil microbial biomass C, but the SIRs for D-glutamine and D-glutamic acid were not constant fractions of the corresponding L-amino acid SIRs. Despite the sensitivity of D-amino acid SIR to streptomycin, the D-amino acid SIR did not provide the basis for distinguishing between bacterial and total microbial activity in soil. The ratio of L-amino acid SIR to D-amino acid SIR (L:D ratio) declined sharply as the metabolic quotient (qCO_2; respiration rate divided by biomass) increased.

In this paper we discuss chirality as a factor in amino acid utilization by soil microbial communities, attempt to put our observations on this subject into their historical context and indicate how the theoretical framework for chirality and biological activity might be applied to these observations in order to improve characterization of microbial communities in terrestrial ecosystems.

Keywords. Enantiomerism, D-amino acids, L-amino acids, microbial communities, respiration, soil, substrate utilization

1. Introduction

1.1 Terminology and Background

An explanation of the terms used to describe stereoisomerism, in general, and chirality, in particular, is appropriate because some terms are used almost interchangeably and not always correctly; We are not without blame. *Stereoisomerism* refers to molecules having identical atomic composition and identical connectivity, but different spatial arrangements of the atoms. There are two types of stereoisomerism: *enantiomerism* or *chirality* on one hand, and *diastereomerism* or *diastereoisomerism* on the other. This paper is concerned exclusively with chirality. Chirality describes the arrangement of a molecule in space such that it cannot be superimposed on its mirror image, and can be regarded as *handedness*. To be chiral a molecule must have a chiral centre, a chiral plane, a chiral axis or be helical. In amino acids (except glycine which is achiral), the asymmetric arrangement of groups bonded to the α-carbon (the chiral centre) confers on the molecule the property of being non-superimposable on its mirror image.

The ability to rotate the plane of polarised light is associated with chirality, with the enantiomers rotating the plane in different directions, and this is the source of the term *optical isomer* used to describe an enantiomer. A mixture of both enantiomers of a compound is known as a *racemate* or *racemic mixture* and if the enantiomers are present at equimolar concentrations the mixture does not display overall optical activity.

Unambiguous designation of enantiomers is of critical importance and if possible the absolute configuration of enantiomers should be used. In the case of amino acids and sugars, the D- and L- notation is conventionally used. This system of notation uses the configurations of D- and L-2,3-dihydroxypropanal (D- and L-glyceraldehyde) as the conventional references.

The R/S notation has been developed to allow unequivocal assignment of enantiomers. It is imagined that the molecule is being viewed along the axis from the chiral centre to the group with lowest priority in the chemical sequence rules (S > N > C > H, and alkyls > carbonyls), *i.e.* along the chiral C to H axis. If the order of decreasing priority of the other three groups is clockwise, the molecule is designated R (*rectus*) and if it is counter-clockwise, the molecule is designated S (*sinister*). According to the R/S system, all the D-amino acids have the R configuration and all the L-amino acids have the S configuration, with the exception of cysteine for which the bonding of sulphur at the chiral centre means that L-cysteine has the R configuration.

The direction in which polarised light is rotated is a physical property that may be solvent- and wavelength-dependent and independent of absolute configuration for different molecules. The notations based on optical rotation are (+) and (-) to indicate *dextro-* and *levo-*rotation, respectively. The prefixes *d-* and *l-* should be avoided since they can be confused with D- and L-, with which they are not necessarily synonymous.

1.2 Historical context

The first observations leading to the present basis of stereochemistry were made in 1848 by Louis Pasteur, who is better known for his contributions to microbiology than to chemistry (Pasteur, 1860). Prior to this time it was known that some compounds could rotate the plane of polarised light, but the basis of such observations were unknown. Pasteur's discoveries that the two forms of tartaric acid formed crystals which were mirror images and that each rotated the plane of polarised light in opposite directions were serendipitous for two reasons. Firstly, he examined sodium ammonium tartrate, which is virtually the only tartrate to crystallize in mirror images that can be seen and separated mechanically from each other using a contemporary microscope. Secondly, crystallization of this salt into two separate forms only occurs below 26°C; the cool Parisian climate had a role (Roberts, 1989). Pasteur modestly expressed his good fortune: *Dans les champs de l'observation, le hasard ne favorise que les esprits prepares*. The complete explanation that the atoms in different enantiomers were arranged differently in space was provided soon after by van t'Hoff (1874), LeBel (1874) and Wislicenus (1887).

1.3 Chirality in biology

There is a large volume of literature, particularly related to pharmacology, considering the fact that enantiomers interact differently with biological systems, which are themselves chiral and made of chiral constituents (see Beckett, 1959; Testa, 1990; Holmstedt, 1990 for reviews). When the observation that only one enantiomer of asparagine was sweet to taste was reported by Piutti (1886), Pasteur is said to have commented that the nervous tissue might itself be dissymmetric (Holmstedt, 1990). This is thought to be the first mention of chirality and biological activity. When Emil Fischer (1894) outlined the "lock and key" model for the interaction of a reactant at the active-site of an enzyme, he also proposed that enzyme-substrate interactions were enantiospecific and later predicted that such specificity was due to the chirality of enzymes (Fischer, 1898). Since then many reports of enantioselectivity at taste, olfactory and other nervous receptors as well as toxicological, nutritional and pharmacological enantioselectivity have been made (Holmstedt, 1990; Brückner and Hausch, 1990).

In soils, it has been observed that D-amino acids accumulate relative to L-amino acids during the decomposition of organic matter (Kawaguchi et al., 1976); that L-glutamine repressed microbial urease production whereas D-glutamine did not (McCarty et al., 1992); and that glutamine synthetase activity is differentially inhibited by enantiomers of methionine sulphoximine (L. Badalucco, personal communication).

The natural biological occurrence of D-amino acids is restricted to a few specific roles, such as the cross-linking of bacterial peptidoglycan (Weidel and Pelzer, 1964) and some microbially-produced peptide antibiotics e.g. gramicidin and actinomycin D (Kuhn and Somerville, 1971). By far the commonest naturally occurring amino acids are D-alanine and, to a lesser extent D-glutamic acid, both of which are found in bacterial cell walls. Occasionally, other D-amino acids are found in the cell walls, e.g. D-ornithine in Type IV peptides of peptidoglycan (Tipper, 1973). The scarcity of D-amino acids in biological systems and the specific occurrence in bacteria has led to D-alanine being used as an indicator of bacteria in soil microbiological studies (Tunlid and White, 1992; Gunnarson and Tunlid, 1986). However, the biological significance of naturally occurring D-amino acids has more far-reaching implications since it has been argued that the preferential sorption of some L-amino acids and D-glucose over that of the other enantiomers to clays (Bondy and Harrington, 1979) offers a clue to the biochemical origin of the first living cells (Cairns-Smith, 1982).

2. Enantioselective substrate utilization by soil microorganisms and microbial communities

The only references earlier than Hopkins and Ferguson (1994) and Hopkins *et al.* (1994) we have found on enantioselective substrate utilization by soil microorganisms are in Waksman (1932): *Of the optically active amino acids, both forms are attacked in varying proportions by bacteria; in the case of glutamic acid, the rate of decomposition is the same (Neuberg, 1909).* These observations are as tantalising as they are interesting. On the one hand, they clearly indicate that Selman Waksman considered chirality as a relevant factor at a fairly early stage both in the appreciation of the importance of chirality in biology and in the study of soil microbiology. On the other hand, Waksman (1932) did not say which amino acid enantiomer was preferentially used! Moreover, we now question the veracity of the claim that D- and L-glutamic acid are decomposed at the same rate (Hopkins *et al.*, 1994; Hopkins *et al.*, 1997; O'Dowd *et al.*, 1997).

Many of the compounds used in substrate utilization assays, such as the BIOLOG^R system, are chiral. In some cases both enantiomers of a particular compound are represented either singly or as the racemate, *e.g.* D- and L-alanine and D- and L-serine all occur separately, and racemates of lactic acid, carnitine and α-glycerol phosphate are included in BIOLOG^R GN plates. However, chirality as a factor in substrate utilization assays of microbial communities in environmental samples has not been widely examined.

2.1 Utilization of amino acid enantiomers

For soils, the substrate induced respiration for D-amino acids is significantly less than that for the corresponding L-amino acids. This has now been shown for over 20 soils and with representatives of all the main structural groups of amino acids (Hopkins and Ferguson, 1994; Hopkins *et al.*, 1994; 1997; Hopkins, 1996; O'Dowd *et al.*, 1997) and a similar enantiomeric effect has been shown for D- and L-glutamine induced ammonification in soil (Hopkins, 1996). The data in Fig. 1 are typical of the respiratory responses by soil microbial communities to amino acid enantiomers.

Following the observation of the enantiomeric effect on amino acid SIR, two explanations, that were not necessarily mutually exclusive, were proposed. Either the D-amino acids were utilized by a smaller fraction of the soil microbial community than the corresponding L-amino acids, or that both enantiomers were utilized by a similar fraction of the microbial community but at different rates (Hopkins and Ferguson, 1994). The natural occurrence of D-amino acids in bacteria and the fact that D-amino acid SIR was apparently sensitive to streptomycin and insensitive to cycloheximide, led to the hypothesis that D-amino acid SIR was a function of the bacterial biomass (Hopkins and Ferguson, 1994). This was supported by the observation that the L-amino acid SIR was correlated with soil microbial biomass C whereas the D-amino acid SIR was not a constant fraction of the L-amino acid SIR across a range of soils (Fig. 2).

However, the hypothesis was not supported when tested using two soils with similar total microbial biomass contents (0.42 and 0.53 mg biomass C g^{-1} soil), but different fungal biomass contents (0.16 and 4.4 µg ergosterol g^{-1} soil, respectively) (O'Dowd *et al.*, 1997). Furthermore, L-amino acid SIR is, like D-amino acid SIR, more sensitive to streptomycin than cycloheximide (O'Dowd *et al.*, 1997). It has also been shown that results from streptomycin and cycloheximide used in conjunction with added substrates were unreliable because of breakdown of the antibiotic (Badalucco *et al.*, 1994; Hopkins *et al.*, 1994).

Preincubating soil for 7 days with D-glutamine led to large increases in the subsequent D- and L-glutamine SIR, whereas preincubation with L-glutamine led to a large increase in the subsequent L-amino SIR but a much smaller increase in the subsequent D-glutamic acid SIR (Table 1). This is clearly demonstrated by the effect of preincubation with D- or L-glutamine on the subsequent L:D ratio for glutamic acid (Table 1), and indicates enantiospecificity in D-amino acid utilization.

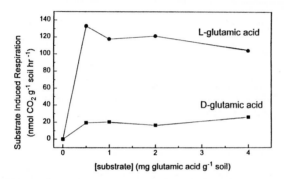

Fig. 1. D- and L-glutamic acid induced respiratory responses in an arable soil. Each point is the mean of three replicates.

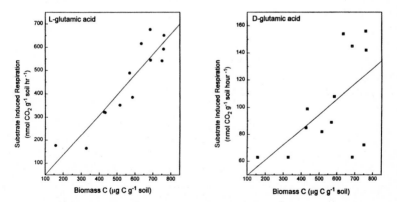

Fig. 2. Relationships between D- or L-glutamic acid SIRs and microbial biomass C determined by glucose SIR for 13 soils. Adapted from Hopkins *et al.* (1997).

Table 1. Effect of preincubation for 7 days at 22°C with 2.0 mg D- or L-glutamic acid g^{-1} soil on the subsequent D- and L-glutamic acid SIRs in an arable soil. Each value is the mean of three replicates and standard deviations are given in brackets.

Preincubation substrate	Substrate for SIR	SIR (nmol CO_2 g^{-1} soil $hour^{-1}$)	L:D ratio
None	D-glutamic acid	15 (3.8)	5.0
None	L-glutamic acid	75 (5.7)	
D-glutamic acid	D-glutamic acid	330 (14.6)	1.6
D-glutamic acid	L-glutamic acid	530 (6.5)	
L-glutamic acid	D-glutamic acid	40 (5.8)	15
L-glutamic acid	L-glutamic acid	600 (28.3)	

Similar effects to those reported in Table 1 were observed by Hopkins and Ferguson (1994) for alanine and glutamine. In general, preincubation with a D-amino acid enhanced both the subsequent D- and L-amino acid SIRs, whereas preincubation with the L-amino acid increased the subsequent L-amino SIR to a far greater extent than the subsequent D-amino acid SIR. Furthermore, there is some evidence in Hopkins and Ferguson (1994) that for D-amino acids the enhancement effect of preincubation showed enantiospecificity, *i.e.* preincubation with D-alanine enhanced D-glutamine SIR (Table 2).

Table 2. Effect of preincubation for 7 days at 22°C with 2.0 mg D- or L-amino acid g^{-1} soil on the subsequent D- and L-amino acid SIRs in a grassland soil. Each value is the mean of three replicates and the standard deviations are given in brackets. Adapted from Hopkins and Ferguson (1994).

Preincubation substrate	Substrate for SIR	SIR (nmol CO_2 g^{-1} soil $hour^{-1}$)
None	D-alanine	85 (8.6)
D-alanine	D-alanine	200 (20.7)
D-glutamine	D-alanine	136 (4.6)
L-alanine	D-alanine	154 (15.6)
L-glutamine	D-alanine	223 (26.0)

The ultimate reason for these enhancement effects is not known, but D-amino acid specific enzyme induction is one possibility and substrate-mediated selection of the microbial community is another. The first of these possibilities implies a change in particular activities levels and the second a change in the composition of the microbial community. In general, the greater the amount of readily available substrate (L-glutamic acid in the case of the data in Table 1), the more enantioselective the microbial community, whereas in the presence of a less readily available substrate (D-glutamic acid in the case of the data in Table 1), the less enantioselective the microbial community. This observation is consistent with the reduction in nutritional fastidiousness of microbial communities under increasingly oligotrophic conditions (Fry, 1990). Comparison of the D- and L-amino acid SIRs for related soils with similar pH but markedly different long-term nutritional inputs (Hopkins et al., 1997) reveals the same effect (Table 3). In the soil that had received large regular applications of farm yard manure the L:D ratios were significantly greater than the soil that had received no organic manure (FYM) for nearly 100 years (Table 3).

The 13 soils considered by Hopkins et al. (1997) had received a wide range of long-term inorganic and organic fertilizer treatments and had markedly different chemical and biological properties (Hopkins and Shiel, 1996). The L:D ratios for glutamine and glutamic acid of these soils declined sharply with increasing qCO_2 (basal respiration/biomass C) (Hopkins et al., 1994). A similar relationship of declining L:D ratio for glutamine and glutamic acid with increasing qCO_2 occurred in the five different soils examined by Hopkins et al. (1994) (Fig. 3).

Table 3. D- and L-amino acid SIRs for soils from long-term experimental grassland plots that had received either 20 t farm yard manure (FYM) ha^{-1} every second year or no manure or fertilizer applications (nil) for nearly 100 years. The FYM soil had pH 5.1 and the nil soil had pH 5.0. Each value is the mean of three replicates and the standard deviations are shown in brackets. Adapted from Hopkins et al. (1997).

Substrate	Substrate induced respiration (nmol CO_2 g^{-1} soil hour^{-1})		L:D ratio	
	FYM	nil	FYM	nil
L-alanine	257 (15.7)	179 (33.1)		
D-alanine	118 (10.6)	121 (8.1)	2.2	1.5
L-glutamine	577 (92.3)	247 (18.2)		
D-glutamine	117 (6.7)	127 (12.7)	4.9	1.9
L-glutamic acid	651 (56.7)	323 (35.0)		
D-glutamic acid	142 (5.3)	85 (2.5)	4.6	3.8

Measurements of qCO_2 are used to infer the physiological status of the soil microbial community (Anderson, 1994), with high qCO_2 being associated with physiologically stressed communities in which the organisms have to expend a relatively large amount of energy per unit of biomass (Nannipieri et al., 1990). If this is the case, the L:D ratios for glutamine and glutamic acid could be an indicator of physiological stress at the community level. The source of the physiological stress in the data from Hopkins et al. (1994) is unclear, but in the data of Hopkins et al. (1997) soil acidity is likely to have been associated with the high qCO_2. It is possible that the organisms in the more stressed communities discriminated between amino acid enantiomers to a lesser extent than those in the less stressed communities, in response to greater demand for respiratory substrates.

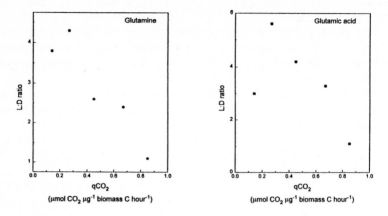

Fig. 3. Relationships between glutamine and glutamic acid L:D ratios and qCO_2 for five different soils. Adapted from Hopkins *et al.* (1994).

2.2 Reactions and regulation of D-amino acid metabolism in microorganisms

The ability to metabolise D-amino acids exists in a range of microorganisms. For example, Kuhn and Somerville (1971) showed that *Escherichia coli* K-12 utilized a range of D-amino acids in the absence of the corresponding L-enantiomer. D-amino acid oxidases, which are involved in racemization (Meister, 1965), have been detected in some fungi, including *Aspergillus niger*, *Penicillium chrysogenum*, *Penicillium notatum*, *Penicillium roquefortii* and *Neurospora crassa* (Horowitz, 1944; Emerson *et al.*, 1950). The occurrence of fungal D-amino acid oxidases supports the suggestion that the difference between D- and L-amino acid SIR in soils is not due to gross differences in the community composition. The key reactions involved in D-alanine metabolism are summarised in Fig. 4.

L-alanine may be produced by transamination or racemization of D-alanine, however, racemization is the only known route for D-alanine production (Reitzer and Magasanik, 1987). One of the alanine racemases in *Salmonella typhimurium* and *Escherichia coli* is constitutively expressed and has a biosynthetic function, and another is inducible by either L- or D-alanine and has a catabolic function (Wasserman *et al.*, 1983; Wild *et al.*, 1985). The work of Wasserman *et al.* (1983) and Wild *et al.* (1985) is consistent with the observations of D-alanine metabolism and its apparent induction in soils by preincubation of the soil with D-alanine (Table 2). Tebbe and Reber (1991) propose that the first step in the breakdown of the D-enantiomer of the herbicide phosphinothricin (homoalanine-4-methylphosphinic acid; an amino acid analogue) is racemization, rather than deamination. Whether this mechanism is common to the metabolism of D-amino acids has yet to be tested, but the slower rate of D-amino acids SIR and the enhancement effects discussed above could be due to a 'lag' period during racemase induction.

D-alanine may induce L-alanine dehydrogenase, which suggests that preincubation with D-alanine may account in part at least for the enhancement of L-alanine metabolism in soil, following preincubation with D-alanine (Hopkins and Ferguson, 1994).

D-alanine may be one of the substrates for a broad-specificity D-amino acid dehydrogenase (Wild *et al.*, 1985), which is consistent with the enhancement of D-alanine SIR by preincubation with other D-amino acids.

D-alanyl-D-alanine ligase catalyses the conversion of D-alanine to peptidoglycan via a primitive polymerization mechanism (Tipper, 1973), this peptidoglycan is then incorporated into the bacterial cell membrane.

D-alanine obviously has a central role in the metabolism, growth and division of bacterial cells. Fig. 4 shows only some of the more prominent metabolic activities involving the D-isomer of alanine, however there are many other reactions known to occur. An interesting task would be to determine the final partitioning of D-alanine uptaken by bacterial cells *i.e.* the relative proportions of the amino acid used in general exergonic cellular metabolism, in cell wall synthesis for growth and in controlled cell wall autolysis during cell division.

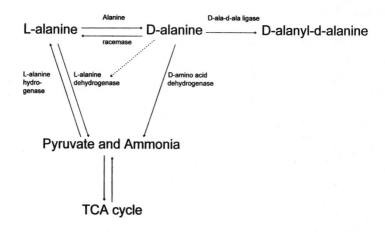

Fig. 4. Summary of the key reactions and regulation of D-alanine metabolism by microorganisms

3. Theoretical considerations

The importance of chirality in biological activity has been considered extensively in pharmacological studies (Cushny, 1926; Beckett, 1959; Portoghese, 1970). From this work a number of generalisations have emerged (see Testa, 1990 for a more detailed discussion). In this section, we relate the theoretical framework developed for chirality and biological activity in pharmacology to the observations of D- and L-amino acid metabolism by soil microbial communities.

3.1 Enantioselectivity and enzyme catalysis

Two distinct types of enantioselectivity by enzymes are recognised (Testa, 1990). *Product enantioselectivity* is the differential formation of two enantiomeric products from an achiral substrate. *Substrate enantioselectivity* is the differential metabolism of two enantiomeric substrates under identical conditions and is the more relevant type. Analysis of the Michaelis-Menten kinetics of enantiomeric substrate utilization has proved useful in pharmacological studies (Testa, 1990; Testa and Mayer, 1988). Three types of substrate enantioselectivity have been identified: enantiomers with similar K_m but different V_{max} values (catalysis-mediated enantioselectivity), enantiomers with different K_m but similar V_{max} (binding-mediated enantioselectivity), and cases where enantiomers display different K_m and V_{max}. In all the cases discussed above differences in V_{max} have been detected, implying catalysis-mediated enantioselectivity. However, we have not examined K_m values for the utilization of amino acid enantiomers by soil microbial communities and cannot exclude the possibility of binding-mediated enantioselectivity. Enantioselective sorption of amino acids to colloids (Bondy and Harrington, 1979) may, however, confound simple interpretation of K_m values, and it is primarily for this reason that we have concentrated on the V_{max} values.

3.2 Pfeiffer's rule and eudismic analysis

Pfeiffer (1956) was the first to report that the greater the biological activity of a racemate, the larger the ratio of the activities of the separate enantiomers. The importance of this observation is that the amount of racemate required to have a particular biological effect must be increased as the relative concentration of the more active enantiomer declines. We have yet to test Pfeiffer's rule with D- and L-amino acids added simultaneously to soils, but suggest that the soil microbial community will offer a major challenge. In pharmacological studies, Pfeiffer's rule is interpreted in terms of the activity of the racemate at a single specific receptor in a single species. Taxonomically and physiologically diverse microbial communities in soils are unlikely to conform to the single receptor model. We have shown that the SIR from soils amended with both D- and L-enantiomers of alanine, glutamine and glutamic acid, at concentrations that would be saturating if added separately, exceeds that of the amino acid enantiomers added separately (Table 4). This implies the existence of 'receptors' for both enantiomers in soil microbial communities. However, the data in Table 4 cannot be used to refute unequivocally Pfeiffer's rule for soil microbial communities because the amino acids were present at saturating concentrations.

Lehmann and others (Lehmann, 1986, 1987; Lehmann et al., 1976) described the more and less biologically-active enantiomers as the *eutomer* and the affinities or efficiencies as substrates) is called the *eudismic ratio* (ER) and its

Table 4. Substrate induced respiration (SIR) for an arable soil amended with saturating concentrations (2.0 mg g^{-1} soil) of D- and/or L-alanine. The values are the means of three replicates and the standard deviations are shown in brackets. Adapted from O'Dowd *et al.* (in press).

Substrate	SIR (nmol CO_2 g^{-1} soil hour^{-1})
D-alanine	39 (4.9)
L-alanine + D-alanine	65 (3.0)
L-alanine	56 (5.8)

logarithm is the *eudismic index* (EI). It is generally found for a series of enantiomeric analogues that EI is positively correlated with both eutomer activity and affinity. In the case of affinity, the slope is referred to as the *eudismic affinity quotient* (EAQ) and represents the increase in chiral discrimination per unit increase in affinity, and can be taken as a quantitative measure of the binding enantioselectivity displayed by a receptor towards a series of stereoisomeric ligands (Testa, 1990). Clearly, testing this relationship for soil microbial communities is required. If soil microbial communities conform, the possibility exists for comparison between soils.

4. Concluding remarks

The D-enantiomers of amino acids are metabolised by some or all soil microorganisms. Chirality in substrate utilization assays provides an approach to investigating the differences in metabolic activity of soil microbial communities. It is clear that the metabolism of D- and L-amino acid enantiomers varies between soils. It is equally clear that this difference does not provide a simple approach to partitioning the soil microbial community since it appears closely related to microbial activity and does not give results consistent with SIR measurements for soils with different bacterial:fungal ratios. At present we have been able to relate the difference in D- and L-amino acid metabolism to the overall physiological state of the soil microbial community, with the organisms exhibiting less catalysis-mediated enantioselectivity with increasing qCO$_2$. This effect may be related to nutritionally- and acidity-induced stress or a range of other environmental factors.

5. Acknowledgements

We are grateful to Kenny McConkey for his contribution to the data in Fig. 1 and Table 1. We acknowledge useful discussions with Luigi Badalucco, Paolo Nannipieri and Declan Barraclough during the preparation of this paper. We are grateful to the Department of Agriculture for Northern Ireland and the UK

226

Biotechnology and Biological Sciences Research Council for their respective contributions to a studentship for RWO.

6. References

Anderson T-H (1994) Physiological analysis of microbial communities in soil: applications and limitations. In: Ritz K, Dighton J, Giller KE (eds) Beyond the Biomass: Compositional and Functional Analysis of Soil Microbial Communities, John Wiley and Sons, Chichester, pp. 67-76

Badalucco L, Pomare F, Grego S, Landi L, Nannipieri P (1994) Activity and biodegradation of streptomycin and cycloheximide in soil. Biol Fertil Soils 18: 334-340

Beckett AH (1959) Stereochemical factors in biological activity. In: Jucker E (ed) Progress in Drug Research, volume 1. Birkhäuser Verlag, Basel, pp. 455-530

Bondy SC, Harrington ME (1979) L-amino acids and D-glucose bind stereospecifically to a colloidal clay. Science 203: 705-711

Brückner H, Hausch M (1990) D-amino acids in food: detection and nutritional aspects. In: Homstedt B, Frank H, Testa B (eds) Chirality and Biological Activity. Alan R. Liss, Inc., New York, pp. 129-136

Cairns-Smith AG (1982) Genetic Takeover and the Mineral Origins of Life. Cambridge University Press, Cambridge.

Cushny AR (1926) Biological Relations of Optically Isomeric Substances. Williams and Wilkins, Baltimore

Emerson RL, Puziss M, Knight SG (1950) The D-amino acid oxidase of molds. Arch Biochem 25: 299-308

Fischer E (1894) Einfluss der Konfiguration auf die Wirkung der Enzyme. Ber Dtsch Chem Ges 27: 2985-2993

Fischer E (1898) Bedeutung der Steriochemie für die Physiologie. Zeitschr Physiol Chem 26: 60

Fry J (1990) Oligotrophy. In: Edwards C (ed) Microbiology of Extreme Environments, Open University Press, Milton Keynes, UK, pp. 93-116

Gunnarson T, Tunlid A (1986) Recycling of fecal pellets in isopods: microorganisms and nitrogen compounds as potential food for Oniscus asellus L. Soil Biol Biochem 18: 595-600

Holmstedt B (1990) The use of enantiomers in biological studies: An historical review. In: Holmstedt B, Frank H and Testa B (eds) Chirality and Biological Activity, Alan R. Liss, Inc, New York, pp. 1-14

Hopkins DW (1996) D-amino acid metabolism in soils. In: van Cleemput O, Vermoesen A, Hofman G (eds) Progress in Nitrogen Cycling Studies, Kluwer Publishers, Dordrecht, pp. 57-61

Hopkins DW, Ferguson KE (1994) Substrate induced respiration in soil amended with different amino acid isomers. Appl Soil Ecol 1: 75-81

Hopkins DW, Isabella BL, Scott SE (1994) Relationship between microbial biomass and substrate induced respiration in soils amended with D- and L-isomers of amino acids. Soil Biol Biochem 26: 1623-1627

Hopkins DW, O'Dowd RW, Shiel RS (1997) Comparison of D- and L-amino acid metabolism in soils with differing microbial biomass and activity. Soil Biol Biochem - in press

Hopkins DW, Shiel RS (1996) Size and activity of soil microbial communities in long-term experimental grassland plots treated with manure and inorganic fertilizers. Biol Fertil Soils 22: 66-70

Horowitz NH (1944) The D-amino acid oxidase of *Neurospora*. J Biol Chem 154: 141-149

Kawaguchi SJ, Kai H, Harada T (1976) The occurrence of D-amino acids in soil organic matter and significance in soil nitrogen metabolism. Soil Sci Plant Nutr 23: 103-105

Kuhn J, Somerville R (1971) Mutant strains of *Escherichia coli* that use D-amino acids. Proc Natl Acad Sci (Washington) 68: 2484-2487

LeBel JA (1874) On the relations which exist between the atomic formulas of organic compounds and the rotatory power of their solutions. In: Memoirs on Stereochemistry, pp. 47-59

Lehmann FP (1986) Stereoisomerism and drug action. Trend Pharm Sci 7: 281-285

Lehmann FP (1987) A quantitative stereostructure activity relationship analysis of the binding of promiscuous chiral ligands to different receptors. Quant Struct Act Relat 6: 57-65

Lehmann FP, Rodrigues de Miranda JF, Ariëns EJ (1976) Stereoselectivity and affinity in molecular pharmacology. In: Jucker E (ed) Progress in Drug Research vol 20, Birkhäuser Verlag, Basel, pp. 102-142

McCarty GW, Shogren DR, Bremner JM (1992) Regulation of urease production in soil by microbial assimilation of nitrogen. Biol Fertil Soils 12: 261-264

Meister A (1965) Biochemistry of the amino acids vol 1. Academic Press, London.

Nannipieri P, Grego S, Ceccanti B (1990) Ecological significance of the biological activity in soil. In: J-M. Bollag and G. Stotyzky (eds) Soil Biochemistry vol 6, Marcel Dekker, New York, pp. 293-355

Neuberg C (1909) Biochem Zeitschr 18: 431-434. Cited in Waksman (1932)

O'Dowd RW, Parsons R, Hopkins DW (1997) Soil respiration induced by the D- and L-isomers of a range of amino acids. Soil Biol Biochem -in press

Pasteur L (1860) On the asymmetry of naturally occurring organic compounds. (Two lectures delivered before the Chemical Society of Paris, 20 Jan and 3 Feb 1860). In: Memoirs on Stereochemistry, pp. 1-33.

Pfeiffer CC (1956) Optical isomerism and pharmacological activity, a generalisation. Science 124: 29-31

Piutti MA (1886) Sur une novelle espèce d'asparagine. Comp Rend Hedb Séances Acad Sci 134-138

Portoghese PS (1970) Relationship between stereostructure and pharmacological activities. Ann Rev Pharm 10: 51-76

228

Reitzer LJ, Magasanik B (1987) Ammonia assimilation and the biosythesis of glutamine, glutamate, aspartate, asaparagine, L-alanine and D-alanine. In: Neidhardt C (ed) *Escherichia coli* and *Salmonella typhimurium* - Cellular and Molecular Biology, American Society of Microbiology, Washington, pp. 302-320

Roberts RM (1989) Serendipity - Accidental Discoveries in Science. John Wiley and Sons, New York.

Tebbe CC, Reber HH (1991) Degradation of [^{14}C]phosphinothricin (glufosinate) in soil under laboratory conditions: Effects of concentration and soil amendments on $^{14}CO_2$ production. Biol Fertil Soils 11: 62-67

Testa B (1990) Definitions and concepts in biochirality. In: Holmstedt B, Frank H, Testa B (eds) Chirality and Biological Activity, Alan R. Liss, Inc., New York, pp. 15-32

Testa B, Mayer JM (1988) Stereoselective drug metabolism and its significance in drug research. In: E. Jucker (ed) Progress in Drug Research vol 32, Birkhäuser Verlag, Basel, pp. 249-313

Tipper DJ (1973) Bacterial cell walls. In: Fasman GD, Timasheff SN (eds) Subunits in Biological Systems part 2, Marcel Dekker Inc, New York, pp, 121-205

Tunlid A, White DC (1992) Biochemical analysis of biomass, community structure, nutritional status and metabolic activity of microbial communities in soil. In: Stotzky G, Bollag J-M (eds) Soil Biochemistry vol 7, Marcel Dekker, New York, pp. 229-262

van t'Hoff JH (1874) A suggestion looking to the extension into space of the structural formulas at present used in chemistry. And a note upon the relation between the optical activity and the chemical constitution of organic compounds. In: Memoirs on Stereochemistry, pp. 35-45

Waksman SA (1932) Principles of Soil Microbiology, 2nd edition. Baillière, Tindall and Cox, London

Wasserman SA, Walsh CT Botstein D (1983) Two alanine racemase genes in *Salmonella typhimurium* that differ in structure and function. J Bact 153: 1439-1450

Weidel W, Pelzer H (1964) Bag-shaped macromolecules - a new outlook on bacterial cell walls. Adv Enzymol 26: 193-232

Wild J, Hennig J, Lobocka M, Walczak W, Klopotowski T (1985) Identification of the *dadX* gene coding for the predominant isozyme of alanine racemase in *Escherichia coli* K12. Mol Gen Genet 198: 315-322

Wislicenus J (1887) The space arrangement of the atoms in organic compounds and the resulting geometrical isomerism in unsaturated compounds. In: Memoirs on Stereochemistry, pp. 61-131

Analysis of the Bacterial Community according to Colony Development on Solid Medium

Tsutomu Hattori[1], Hisayuki Mitsui[1], Reiko Hattori[2], Shuichi Shikano[3], Krystyna Gorlach[1], Yasuhiro Kasahara[1] and Adel El-Beltagy[1]

[1]Institute of Genetic Ecology, Tohoku University, Katahita 2-1-1, [2]Attic Laboratory, Komegafukuro 1-6-2-401, [3]Department of Biology, Faculty of Science, Tohok University, Aramaki; Sendai 980, Japan

Abstract. Bacterial communities of paddy and grassland soils were studied by a plate count method. The colony formation processes were simulated by a superposition of the first order reaction (FOR) model curves. Isolates from colonies formed along each FOR curve were collected as a colony forming curve (CFC) group. Sequencing 16SrDNA base sequences revealed that isolates had diverse phylogenetic positions and those of each CFC group showed a trend to make clusters. Bacteria in microaggregate clumps (ca. 0.1 mg) also showed CFC groups similar to those obtained with 1g of soil and were phylogenetically diverse. Based on the results we propose to catalogue all the bacteria of the community in the microaggregates.

Keywords. Plate count, microbial community, oligotrophic bacteria, phylogenetic cluster, soil microaggregates, bacterial diversity

1. Introduction

A large number of bacterial species reside in soil. However there is still a paucity of information on their taxonomic or phylogenetic positions. Therefore, the soil bacterial community is often considered as a black box, which is examined only in a limited manner.

Hattori (1996) proposed an idea as a central dogma; soil bacteria residing in the soil have two alternatively convertible forms or states, that is, the active or vegetative state and the quiescent or persistent state. The former are able to divide but the latter can not unless they convert to the active cells: here the terminology of activity and quiescence is limited to the context of the ability of cell division. We may consider that active cells, as well as introduced cells, may not be able to survive for a long time under soil conditions, unless they convert to persistent cells. According to the dogma we consider that the plate count measures the active cells of soil bacteria or the active community.

The present paper addresses the following questions: First, what can we learn about the active community by the plating method ? Second, can we list fully the members of the active and whole soil community ?

2. Materials and methods

Soils. Paddy soil and grassland soils described previously were used (Hashimoto and Hattori 1989, Kasahara and Hattori 1991).

Culture medium. The DNB medium which is a 100-fold dilution of the conventional nutrient broth (NB medium) was used unless described otherwise (Hattori and Hattori 1980).

Plate counting and isolation of bacteria. Unless described otherwise 1g of soil sample taken from a grassland and a paddy field near Sendai was, after being dispersed and serially diluted, incubated on the DNB medium with 1% agar at 27°C. The numbers of small colonies appearing (ca. 0.1 mm diameter) were counted every 3-6 h during the first 48 h and then at longer intervals (12-48 h or more).

Colony forming curves (CFC). The increase in the number of colonies was simulated by a superposition of FOR model curves (component CFC), which was originally presented to describe the formation process on plate of visible colonies N_t of single cell populations as a function of incubation time t (Hattori 1988):

$$N_t = N_\infty \{1 - \exp - \lambda (t - t_r)\} \qquad (t > t_r)$$

where N_∞, λ and t_r are parameters expressing the expected number of colonies at infinite incubation time, the probability of the occurrence of colony formation in a unit time and a retardation time for the appearance of the first colony, respectively (Ishiguri and Hattori 1985). More than 30 organisms were isolated from colonies appearing along each component CFC and collected as a CFC group.

Sequencing of 16S rDNA. Using genomic DNA from each organism, a 0.5 kb 5'-terminal region of 16S rDNA genes was amplified by PCR. The PCR product was sequenced using a 373A DNA sequencer (Applied Biosystems). The sequences were analyzed by the neighbor-joining method and phylogenetic trees were constructed.

3. Results

3.1 Colony-forming curve (CFC) and CFC groups of soil bacteria

First, the kinetics of colony formation of single cell populations was studied. A population of single cells of the same bacterial species on a plate did not form their colonies simultaneously, but did according to the first order reaction kinetics (FOR model). The result was interpreted that the division of bacterial cells on a plate initiate at random similar to the decay of the radio-active atoms (Hattori 1988).

Second, we extended our study to the relationship between the incubation time and the formation of colonies of soil bacteria on plates. The process was simulated by a colony-forming curve (CFC) which is a super-position of FOR model curves (all curves are referred to as the component colony-forming curves, cCFC; Hashimoto and Hattori 1989; Kasahara and Hattori; 1991; Gorlach and Hattori 1994), as shown in Fig. 1.

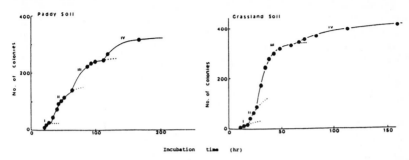

Fig. 1. Colony forming curves for the grassland paddy soils.Numbers of colonies per plate were plotted.

Each CFC can be characterized by a set of t_r values for each cCFCs. Superposing patterns of cCFCs of bacteria in the grassland and the paddy soil were different from each other. The value of t_r is approximated by the time of crossing of each cCFC with the horizontal axis or the preceding cCFC (Fig. 1). The patterns for bacteria in one soil sampled at different times were essentially constant. However a marked difference was observed between the patterns of the paddy and the grassland soils as shown in Fig. 1. The numbers of cCFCs for the paddy and the grassland soils appeared during the first 100 h of incubation were 3 and 4, respectively. Usually additional cCFCs were observed when the incubation time exceeded 100 h (Fig. 1).

To investigate the community structure, we isolated bacteria from colonies formed along each cCFC and collected as a CFC group, which is, respectively, referred to as CFC-I, -II, -III and -IV according to the order of appearance. A collection of bacterial cultures comprising those from all CFC groups is referred to as an ecocollection. Growth rates (expressed by t_r values) of each isolate belonging to each CFC group, respectively, showed concentrated distribution patterns. From these results one may consider that the active bacterial community in a soil consists of several unit groups; in other words, CFC groups are, respectively, the representatives of unit groups of the community.

It is of interest that physiological and physicochemical characters of isolates showed trends to shift from CFC-I to -IV: the percentage of isolates sensitive to the NB medium and 1%NaCl were increased (Hattori 1976). From CFC-1 to -IV the surface negative charge of cells decreased and the hydrophobicity of cell surface increased (Morisaki *et al* 1993, Kasahara *et al*. 1993, Shingaki *et al* 1994).

3.2 Phylogenetic positions of organisms of ecocollection

It is widely believed that only limited kinds of soil bacteria produce colonies, and thus the phylogenetic positions of isolates from the plate will not be diverse but

232

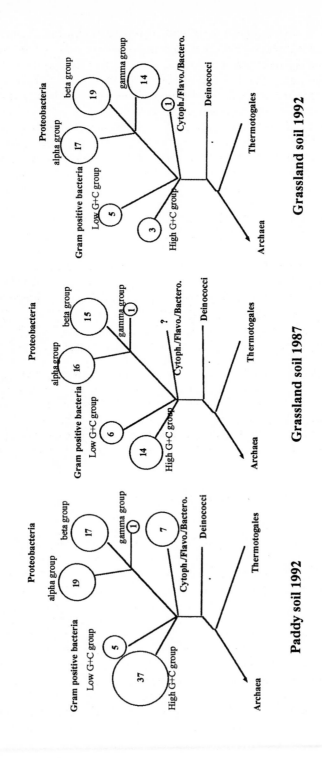

Fig.2. Phylogenetic positions of ecocollection. Numbers of tested strains for ecocollections of paddy soil 1992, grassland soil 1987 and 1992 are, respectively, 86, 32 and 59. Figures in circles indicate the number of isolates which belonged to each phylogenetic group.

limited to only a few lines. However the sequencing of a 0.5 kb of 5'-terminal region of 16S rDNA from organisms of the ecocollections showed that their phylogenetic positions were largely diverse. Isolates of both the paddy and the grassland ecocollections belonged to different subdivisions of eubacteria; Proteobacteria alpha, beta and gamma subgroups, Gram positive-high and -low G+C groups and Cytophaga/Flexibacter/ Bacteroides group (Fig. 2).

No organism had a DNA base sequence which matched perfectly those in the GenBank/EMBL/ DDBJ database, although the DNA of 53% of tested organisms showed silimarity values higher than 95%. Some of the isolates showed identical sequences to each other. However, by further analysis using a 1.4 kb region of 16S rRNA, isolates with the same base sequence taken from the grassland ecocollection 1987 were clearly distinguished from each other and no pair of isolates so far examined showed an identical sequence. Most isolates of each CFC group showed a trend to make phylogenetically-linked clusters. The details of these results will be published elsewhere.

The phylogenetic analysis of ecocollections revealed a notable tendency. Near to the phylogenetic position of *Agromonas oligotrophica* which was isolated from the paddy soil and identified by Ohta and Hattori (1983), many oligotrophic (DNB) organisms of both the paddy and grassland ecocollections took their positions and, furthermore, the positions of *Bradyrhizobium japonicum*, *Blastobacter denitrificans* and *Nitrobacter winogradsky* (and also *Afipia felis*; an animal parasite) were also found there. We can consider that several soil bacterial groups which have respectively different but important ecological roles in the terrestrial environment have a common phylogenetic origin in the region. From this aspect Saitou and Hattori (1995) proposed to call the phylogenetic region comprised of these terrestrial bacteria as the BANA (Bradyrhizobium-Agromonas-Nitrobacter-Afipia) region; the region may be significant only for the terrestrial bacteria or for the understanding the diversification of the terrestrial bacteria.

3.3 Bacteria in 0.1 mg of soil microaggregates clumps

In the investigation of soil bacteria it has been traditional to start plate counting experiments with one gram of soil without any justification. However, if we intend to catalogue all of the active or the whole bacteria by their taxonomic or phylogenetic positions, the number of bacteria residing in this amount of soil is too large. Our previous studies on soil bacteria, fungi, and protozoa (Hattori 1973, Hattori 1988, Vargas and Hattori 1990) showed that soil aggregates are more a significant unit of the soil for microbial ecology; macroaggregates for protozoa and fungi, and microaggregates for bacteria (and some microprotozoa). To examine whether we can study the soil bacterial community with a small amount of soil that individual bacteria in it can be categorized, three microaggregate clumps (0.1-0.4 mg), which were positioned less than several mm apart from each other *in situ* in the grassland soil (1994), were sampled. After dispersion in 10 ml of a sterilized water by sonic oscillation (40W, 1 min) the DNB medium plates were inoculated with 1 ml of the suspension and then incubated. Interestingly, bacteria in these clumps formed colonies depicting own CFC, the patterns of which consisted of four cCFCs and were essentially similar to each other and also to those of the soil samples of 1987 and 1992. The results indicate that bacteria in these small samples have still the same structure as the whole community.

Although we determined the base sequence of 16SrDNA from only 27 isolates (clump A, 9 isolates; clump B, 11 isolates; clump C, 7 isolates) of each CFC-III group, they were phylogenetically very diverse forming phylogenetically linked clusters, PLC (less than 2% difference). We recognized 20 PLCs in the Gram positive high and low G+C groups and the Proteobacteria beta subgroup. Only 2 PLCs contained isolates from two clumps and other PLCs consisted of isolates obtained from only one clump. Other CFC groups of the clumps will contain other phylogenetic groups; bacteria will be very diverse, although they were placed *in situ* very closely; only a few mm from each other. We may consider that bacterial communities in microaggregate clusters positioned closely *in situ* in the grassland field had the same unit structure but different species composition.

4. Discussion

Soil bacteria formed colonies according to a CFC, the pattern of which was specific for each soil. This offers the key to analyze the active bacterial community of the soils. We may consider that all CFC groups are unit ensembles consisting of a bacterial community. Probably among CFC groups there may be some close interractions through signaling molecules which control the alternative conversion of the cell states.

The fact that bacteria within the microaggregate clumps showed CFC similar to bacteria within one gram soil indicates that the relatively small number of bacteria in a clump still contained the same structure as the community. The number of the active cells may be several thousands at most. This figure seems in a range that we will be able to isolate all the bacteria and characterize them by 16SrDNA sequencing and/or other methods. Thus we could know the identity of all the active bacteria at the time of sampling. The species composition of the active community will change from time to time through dynamic equilibrium between the active and quiescent forms. The result of such cataloguing will give basic information on bacterial community structure and biodiversity.

5. References

Gorlach K, Hattori T (1994) Construction of eco-collection of paddy field soil bacteria for population analysis. J Gen Appl Microbiol 40:509-517

Hashimoto T, Hattori T (1989) Grouping of soil bacteria by analysis of colony formation on agar plates. Biol Fertil Soils 7: 198-201

Hattori T (1973) Microbial Life in the Soil. Marcel Dekker, New York, pp 263-312

Hattori T (1976) Plate count of bacteria in soil on a diluted broth as a culture medium. Rep Inst Agr Res Tohoku Univ 27: 23-30

Hattori R, Hattori T (1980) Sensitivity to salts and organic compounds of soil bacteria isolated on diluted media. J Gen Appl Microbiol 26: 1-14

Hattori T (1988) The Viable Count: Quantitative and Environmental Aspect. Science Tech, Madison. Springer Verlag, Berlin, pp. 16-42

Hattori T (1988) Soil aggregates as microhabitats of soil microorganisms. Rep Inst Agr Res Tohoku Univ 37: 23-36

Hattori T (1996) Advances in soil microbial ecology and diversity. In : Samson RA, Stalpers JA, van der Mei D, Stouthamer AH (eds) Culture Collection to Improve the Quality of Life, Centralbureau voor Schimmelcultures, Baarn, The Netherlands. pp. 118-122

Ishiguri S, Hattori T (1985) Formation of bacterial colonies in successive time intervals. Appl Environ Microbiol 49 : 870-873

Kasahara Y, Hattori T (1991) Analysis of bacterial populations in a grassland soil according to rates of development on solid media. FEMS Microbiol Ecol 86 : 95-102

Kasahara Y, Morisaki H, Hattori T (1993) Hydrophobicity of the cells of fast- and slow-growing bacteria isolated from a grassland soil. J Gen Appl Microbiol 39: 381-388

Morisaki H, Kasahara H and Hattori T (1993) The cell surface charge of fast- and slow-growing bacteria isolated from grassland soil. J Gen Appl Microbiol 39: 65-74

Ohta H and Hattori T (1983) *Agromonas oligotrophica* gen. nov., sp. nov., a nitrogen-fixing oligotrophic bacteria. Antonie van Leeuwenhoek 49: 429-446

Saitou A and Hattori T (1995) Classification of grassland DNB organisms belonging to the BANA (Bradyrhizobium, Agromonas, Nitrobacter, Afipia) region of the Proteobacteria alpha subgroup. Abstract of the 11th meeting of Japanese Society of Microbial Ecology at Fukuoka. p. 25

Shingaki R, Gorlach K, Hattori T, Samukawa K and Morisaki H (1994) The cell charge of fast- and slow-growing bacteria isolated from a paddy soil. J Gen Appl Microbiol 40: 469-475

Vargas R and Hattori (1990) The distribution of protozoa among soil aggregates. FEMS Microbiology Ecology 74: 73-78

The Complexity of the Flux of Natural Substrates in Soils: A Freeze-thaw can Increase the Formation of Ischemic or Anaerobic Microsites

A. Zsolnay[1]

[1] Institut für Bodenökologie, GSF-Forschungszentrum für Umwelt und Gesundheit, D 85764 Neuherberg bei München

Abstract. A study was undertaken to determine the flux and ecological impact of naturally occurring substrate as the result of a natural perturbation: a freeze-thaw. For 3 different arable soils an increase in the amount of water extractable organic carbon (WEOC) was found after such an event. The amount of this increase varied with the water content of the soil. The amount of this WEOC, which was relatively labile and usable as a substrate, was between 166 and 447 µmoles C per litre of soil. A freeze-thaw cycle increased this by an additional 181 to 578 µmoles. The material extractable after a freeze-thaw was relatively more labile than that obtained from unfrozen soil. Simple calculations indicate that the metabolism of WEOC, especially after a freeze-thaw, could depress the oxygen content in a soil. Water saturating the soil before freezing did not produce consistent results in regards to the increase of labile soluble carbon.

Keywords. Natural substrate, water extractable organic carbon, DOM, DOC, freeze-thaw, labile, oxygen, anoxic

1. Introduction

Substrate utilisation tests are increasingly used to characterise microbial communities in the agricultural environment. One difficulty is that many of these tests come from the field of medicine and are oriented toward substrates, which are appropriate for this area. However, the application of such tests in environmental studies can theoretically lead to a bias, tending to emphasise the presence of organisms, which may not play the chief ecological role *in situ*. One potential solution to overcome this is the use of „natural" substrates in such tests. However, the selection of substrates for this is not straightforward. It is also complicated by the fact that the substrates, which are available in soil, are by no means constant in either their quantity or quality. That is to say that they presumably change as the result of natural and anthropogenic perturbations. The study here investigated such a dynamic substrate flux caused by a freeze-thaw event and also estimated the possible ecological effect of such a perturbation. Only through understanding the nature and impact of substrates *in situ* can we improve the selection of substrates and techniques used in ecologically relevant substrate utilisation tests.

A major ecological function of the dissolved organic material (DOM) or humus in a soil may be its ability to act as a substrate (McGill *et al.*, 1986; Zsolnay and Steindl, 1991; Zsolnay, 1993). Since it is, by definition, the most mobile abiotic organic soil fraction, it can be present in the soil's microsites, and its metabolism may result in a depletion of oxygen, which can in turn result in denitrification, which can have both a positive (ground water protection) and negative (greenhouse gas production) effect (reviewed by Zsolnay, 1996).

A freeze-thaw event, such as often occurs in the world's temperate zones, has been shown to increase the amount of dissolved organic material (DOM) in agricultural soils (Christensen and Christensen, 1991; Wang and Bettany, 1993). Previous field research (Bremner *et al.*, 1980; Goodroad and Keeney, 1984; Edwards and Killham, 1986; Cates and Keeney, 1987; Christensen and Tiedje, 1990) has shown, but not always (Seiler and Conrad, 1981), that a substantial release of N_2O may occur during a spring thaw. Similarly, laboratory studies indicate that freeze-thaw often (Goodroad and Keeney, 1984; Christensen and Christensen, 1991; Christensen, 1992), but not always (Goodroad and Keeney, 1984; Christensen, 1992), results in a strong increase in denitrification. One important aspect of the effect of freeze-thaw on the soil's DOM, which is absent in the above cited research, is the influence of the soil's water content before the freezing event takes place.

The goals of the reported research were 1) to verify that a freeze-thaw results in a DOM increase and 2) to estimate the labile carbon in DOM, obtained with or without a freeze-thaw cycle.

In addition the effect of water saturating a soil before a freeze-thaw was investigated. This was felt to be of interest, since in many areas during the time of snow melting, it is quite possible that a soil will become water saturated and subsequently be exposed to freezing conditions

The determination of *in situ* DOM is experimentally difficult. Therefore, water extractable matter (WEOM), usually quantified as water extractable carbon (WEOC), is often used as an estimate of the potential amount of DOM in a soil's matrix. This was reviewed by Zsolnay (1996).

2. Methods

2.1 Soils used

Surface soils from three different fields in the proximity of Munich were sampled: a sandy loam (pH 6.5, 7% clay, 26% silt, 67% sand, 1.1% organic carbon) with a water holding capacity (WHC) of 25.9%; a silt loam (pH 6.9, 25% clay, 58% silt, 17% sand, 1.4% organic carbon) with a WHC of 29.7%; and a modified muck soil (pH 5.6, 31% clay, 40% silt, 29% sand, 11.9% organic carbon) with a WHC of 54.3%. The muck soil had been modified in the past by the mixing of a non-muck soil into it to improve its agricultural properties.

2.2 WEOC as a function of freeze-thaw and water content

In the laboratory the field fresh soils were brought to their WHC. Portions of the soils were then allowed to air dry until they had attained the water contents given in the Figs. In addition, one portion was fully air dried while another was oven dried at 105°. WEOC was extracted from an aliquot of each portion (50 g dry weight equivalent) with 100 ml of a 10 mM CaCl$_2$ solution. The extract was then filtered through a 0.4 μm polycarbonate (Nuclepore) filter. The organic carbon in the filtrate was quantified with a high temperature catalytic organic carbon analyser (Shimadsu TOC-5050). All extractions were carried out in triplicate. An aliquot of each portion was frozen at -18° for 7 days in the freezer section of a standard refrigerator. After thawing at room temperature overnight, WEOC was again determined.

2.3 Labile organic carbon determinations

The amount of WEOC, which was removed from solution over a 7 day incubation under ideal conditions, was considered to be labile. The WEOC was obtained as described in 2.2 from (1) field fresh soil at their normal water contents at the time of sampling, (2) from field fresh soils, which had been frozen at -18° C for 4 days and thawed, and (3) from water saturated soils, which had also been frozen and thawed.

The incubation itself was done in the dark at 20° C in Teflon containers, to which 8 ml of WEOC solution and 3 ml of an inorganic nutrient solution had been added. The nutrient solution contained 0.5% of both (NH$_4$)NO$_3$ and K$_2$HPO$_4$. Also added was an inoculum, consisting of 50 μl of unfiltered aqueous soil extract. A blank and a control were also run. In the former the WEOC solution was replaced with 10 mM CaCl$_2$, while in the latter it was replaced with a 1 mM organic carbon glucose solution.

3. Results

3.1 The effect of freeze-thaw on WEOC content

In Fig. 1 the relative changes in WEOC concentrations as a function of freeze-thaw are presented. The maximum effect from freeze-thaw occurred at water contents in the range of 35-56% WHC. Freeze-thaw, presumably, results in an increase in WEOC by expanding the soil's pores, enabling more material to leach out and/or by rupturing a portion of the biomass. One can conjecture that at higher water contents, a greater portion of the soil's water will be outside of the pore space. If this „external" water freezes before the water in the pore space, it could restrict the expansion of the pores. Therefore, the effect of freeze-thaw would be diminished. Below a specific water content, the effect of freeze-thaw decreased as the soils became drier. The extreme case for this was the oven dry soil were no effect as a result of freeze-thaw could be determined.

This research confirms previous work, which has shown that freeze-thaw can result in a significant increase of DOM in soil. However, what has been found here in addition is that this effect can not be generalised. Its magnitude will be controlled by the type of soil and, more importantly, by the water content of the soil before the freezing event. These results should be integrated into the interpretation of any research showing that freeze-thaw has an effect on an ecological or environmental process via DOM.

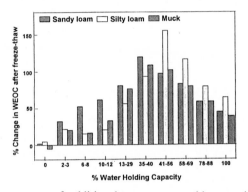

Fig. 1. The amount of additional water extractable organic carbon (WEOC) obtained from 3 soils after a freeze-thaw.

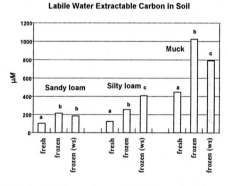

Fig. 2. The amount of labile water extractable carbon in a given volume of soil. „ws" indicates that the soil was water saturated before freezing. Similar letters for a given soil indicate no significant difference at the 95% confidence level.

3.2 Labile WEOC

The amount of labile water extractable carbon is shown in Fig. 2. It can be seen that in the case of all 3 soils, freeze-thaw increased the concentration of this material. The effect of water saturation before a freeze-thaw was variable.

In Fig. 3, it can be seen that the WEOC, which was obtained after a freeze-thaw, was in all cases more labile than the material obtained when the soil was not frozen. Again, the effect of water saturating the soil before the freeze-thaw was variable.

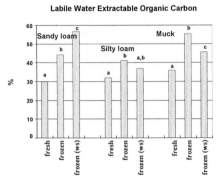

Fig. 3. The relative labile water extractable organic carbon. „ws" indicates that the soil was water saturated before freezing. Similar letters for a given soil indicate no significant difference at the 95% confidence level.

4. Conclusions

The conclusions are summarised in Table 1. It is assumed that the oxygen concentration is at saturation in the pore water and that it is at normal atmospheric levels in the soil's pore space. Both of these assumptions are most likely false, since *in situ* one can expect that a significant proportion of the oxygen has been consumed. Therefore, the given values are most likely too high. On the other hand, the labile WEOC measured is most likely also greater than the *in situ* labile DOC. Under *in situ* conditions, lower temperatures and possibly less nutrients are present. Furthermore, due to the soil's pore structure the availability of DOC to decomposition is presumably less than what was the case during the laboratory incubations. Also, although WEOC is obtained by relatively gentle methods, it probably exceeds the amount of organic matter, which is indeed in the dissolved state within a soil. This in turn is a complicated function of soil pore structure and the amount of available water (Zsolnay, 1996).

Despite all these limitations in estimating the results reported in Table 1, the research presented here does give reason to support the contention that the amount of labile DOM (substrate) in a soil, given proper conditions, has sufficient potential to deplete the oxygen content within the pore water. Furthermore, a freeze-thaw will roughly double this potential.

Acknowledgements. The financial support of the GSF - Forschungszentrum für Umwelt und Gesundheit is gratefully acknowledged.

Table 1. Calculations of potential oxygen content and consumption in 1 litre of soil

	Sandy loam	Silty loam	Muck
density (kg/l)	1.6	1.4	1.0
pore volume (%)	42	47	65
water volume (%)	18	24	40
air volume (%)	24	23	25
O_2 in pore water (μmoles)[a]	45	60	100
O_2 in pore air (μmoles)[b]	2250	2156	2344
labile water extractable organic C (μmoles)	166	177	447
additional labile C (μmoles) after a freeze-thaw	181	185	578
additional labile C (μmoles) after a freeze-thaw[c]	134	397	344

[a]maximum value, assuming saturation
[b]maximum value, assuming 21% O_2 content
[c]soil was water saturated before freezing

5 References

Bremner JM, Robbins SG and Blackmer AM (1980) Seasonal variability in emission of nitrous oxide from soil. Geophys. Res. Letters 7: 641-644.

Cates RL and Keeney DR (1987) Nitrous oxide production throughout the year from fertilized and manured maize fields. J. of Environ. Quality 16: 443-447.

Christensen S and Christensen BT (1991) Organic matter available for denitrification in different soil fractions: effect of freeze-thaw cycles and straw disposal. J. of Soil Sci. 42: 637-647.

Christensen S and Tiedje JM (1990) Brief and vigorous N₂O production by soil at spring thaw. J. of Soil Sci. 41: 1-4.

Christensen S (1992) Decomposability of recalcitrant soil carbon assessed by denitrification. FEMS Microbiology Ecology 82: 267-274.

Edwards AC and Killham K (1986) The effect of freeze-thaw on gaseous nitrogen loss from upland soils. Soil Use and Management 2: 86-91.

Goodroad LL and Keeney DR (1984) Nitrous oxide emissions from soils during thawing. Can. J. Soil Sci. 64: 187-194.

McGill WB, Cannon KR, Robertson JA and Cook FD (1986) Dynamics of soil microbial biomass and water-soluble organic C in Breton L after 50 years of cropping in two rotations. Can. J. Soil Sci. 66: 1-19.

Seiler W and Conrad R (1981) Field measurements of natural and fertilizer induced N₂O release rates from soils. J. Air Pollut. Control Assoc. 31: 767-772.

Wang FL and Bettany JR (1993) Influence of freeze-thaw and flooding on the loss of soluble organic carbon and carbon dioxide from soil. J. of Environ. Quality 22: 709-714.

Zsolnay A and Steindl H (1991) Geovariability and biodegradability of the water-extractable organic material in an agricultural soil. Soil Biology & Biochemistry 23: 1077-1082.

Zsolnay A (1993) The relationship between dissolved organic carbon and basal metabolism in soil. Mitt. d. Österr. Bodenkund. Gesell. 47: 83-95.

Zsolnay A (1996) Dissolved humus in soil waters, In Piccolo A ed: Humic Substances in Terrestrial Ecosystems, Elsevier, Amsterdam, pp. 171-223.

Carbon Transformations During Substrate Utilization by the Microbial Community in an Organic Soil: A Solid-State NMR Study

Elizabeth A. Webster[1,2], J.A. Chudek[2] and D.W. Hopkins[1]

Department of Biological Sciences[1] and Department of Chemistry[2], University of Dundee, Dundee, DD1 4HN, UK

Abstract. We have used cross-polarization (CP) magic angle spinning (MAS) ^{13}C nuclear magnetic resonance spectroscopy to follow the transformations of ^{13}C added to an organic soil as either [13]C_6-D-glucose or [13]C_1-1-L-alanine. After 28 days' incubation, 34% of the added ^{13}C from glucose remained in the soil and using NMR was detected primarily in the alkyl-C and the O-alkyl-C resonances. By contrast, only 12% of the added ^{13}C from alanine remained in the soil after 28 days, and was apparently distributed between a range of functionalities.

Keywords. [13]C, Alanine, glucose, nuclear magnetic resonance spectroscopy, organic soil, substrate utilization

1. Introduction

The abililty to follow the transformations of organic compounds during their utilization by microorganisms in soil is usually restricted by the difficulties of extracting and characterising the products of microbial metabolism. In the present context, the chemical characterisation of the compounds arising during substrate utilization by microorganisms is important because the metabolic products and microbial remains from utilization of one compound are the substrates during subsequent microbial metabolism. Solid-state ^{13}C NMR offers a non-destructive approach to characterising ^{13}C at the level of different functional groups (Table 1) without the need to extract C from the soil (Wilson, 1987). The principal limitations to the use of ^{13}C NMR spectroscopy for soils are the presence of paramagnetic nuclei (Fe, Cu and Mn) in soil and the low ^{13}C natural abundance (1.1 atom%) (Wilson, 1987). We have attempted to overcome these problems by investigating the ^{13}C NMR spectra of an organic soil (containing negligible paramagnetic nuclei) amended with ^{13}C-labelled D-glucose and L-alanine, thereby allowing the transformations of the added ^{13}C to be followed against the background, unlabelled soil C (98.9 atom% of which is ^{12}C and is, therefore, invisible to NMR).

Table 1. Assignments of ^{13}C functional groups to NMR shift ranges. Adapted from Baldock *et al.* (1991) and Hopkins and Chudek (1997).

Shift range (ppm)	Assignment	Typical classes of compounds
0-45	Methyl- & alkyl-C	Aliphatic compounds, lipids, waxes
45-90	O-alkyl- & N-alkyl-C	Polysaccharides, lignin substituents, amino acids, amino sugars
90-110	Acetal- and ketal-C	Polysaccharides
110-160	Aromatic-C	Phenyl-propylene sub-units of lignin
160-200	Carbonyl	Organic acids and peptides

Similar approaches to this have previously been used. Baldock *et al.* (1991) reported that ^{13}C from glucose added to soil was detected predominantly as alkyl-C associated with the clay-sized fractions of soil and as O-alkyl-C associated with the 250-2000 μm size fractions after 34 days incubation. Benzing-Purdie *et al.* (1992) showed that ^{13}C from enriched sodium acetate added to peat was detected as polysaccharide associated with the 0.005-0.05 mm size fraction. Hopkins and Chudek (1997) showed that ^{13}C from enriched *Lolium perenne* leaves was transformed from predominantly polysaccharide-C to alkyl-C-rich compounds during decomposition for 224 days. In the present study, we have compared the utilization of and contrasted the resulting ^{13}C distribution from glucose and alanine in the same soil.

2. Materials and methods

An upland stagnohumic-gley soil from the Redesdale Experimental Husbandry farm, Northumberland, England which is used for rough grazing of sheep was used. The soil contained 45% C, 1.7% total N and had pH 3.1. Further details of the soil biological properties have been provided by Isabella and Hopkins (1994).

2 g samples of soil at 50% water-holding capacity were amended with the equivalent of 2 mg C g^{-1} soil (dry wt) of $^{13}C_6$-D-glucose or $^{13}C_1$-L-alanine (*i.e.* labelled in the methyl-C position), or were left unamended. To ensure even distribution of the substrates in the soil, they were mixed with 0.5 g talc prior to incorporation. The unamended soil contained 410 mmol [13]C g^{-1} soil (equivalent to natural abundance of 1.1 atom%), to which 80 mmol [13]C g^{-1} soil was added as glucose or 54 mmol [13]C g^{-1} soil was added as alanine. The soils were incubated for 28 days at 20°C and their ^{13}C contents were determined in duplicate by mass spectrometry.

Cross-polarization (CP) magic angle spinning (MAS) ^{13}C NMR spectra were recorded with a Chemagnetics CMX LITE 300 MHz spectrometer at 4 kHz MAS, 1 ms contact time and relaxation delay 1 s against tetramethylsilane as the reference. The NMR spectra were divided into the shift ranges based on those in Table 1 and the area under each section of the spectrum estimated using a leaf area meter (ADC Plant Sciences Instrumentation, Hoddeston, Herts, UK). The data from NMR spectra were expressed as the percentage of total spectral area. (See Kinchesh *et al.*, 1995 for a detailed discussion of the difficulties in quantification of NMR spectra.)

3. Results

The amount of ^{13}C remaining in the amended soils declined during incubation, and the ^{13}C loss from alanine-amended soil was greater than that from the glucose amended soil (Table 2).

Fig. 1 shows the spectra for unamended soil, for soils immediately after amendment with either glucose or alanine, and for amended soils after 28 days incubation. At the outset, the effect of ^{13}C-glucose addition was to increase the O-alkyl- and acetal-C signals, with the acetal-C (dioxygenated-C or anomeric-C) from the added glucose appearing as a shoulder at 95 ppm on the larger signal centred at about 100 ppm (attributed to the indigenous acetal-C or ketal-C in the soil). The effect of ^{13}C-alanine addition was to produce a strong signal at 19 ppm characteristic of methyl-C.

Table 2. ^{13}C remaining in the soil at different times. Each value is the mean of two analytical replicates and the standard deviations were less than 3% of the mean.

Time (days)	^{13}C remaining (%)	
	Glucose	Alanine
0	100	100
1	69	22
7	44	15
28	34	12

The data in Table 3 are the percentages of the total spectral area represented by the particular shift ranges (Table 1) at different times. These data provide semi-quantitative indications of the changes in distribution of ^{13}C between the different functional groups with time.

For the glucose-amended soil, the O-alkyl-C area declined from 43% at day 0 to 32% by day 28. The O-alkyl-C signal for the glucose-amended soil after 28 days incubation exceeded that of the O-alkyl-C signal from the unamended soil, indicating that ^{13}C accumulated in O-alkyl-C compounds, other than glucose. The absence of the signal at 95 ppm from the 28 day spectrum for glucose-amended soil, suggests that no unreacted glucose remained in the soil after incubation. In the glucose-amended soil, the methyl- and alkyl-C area was between 24 and 27% throughout the incubation. Since the total ^{13}C remaining in the glucose-amended soil declined to 34% of that added during incubation, the fact that the methyl- and alkyl-C area remained relatively stable indicates accumulation of methyl- and alkyl-C compounds. This suggestion is supported by the greater intensity of the methyl- and alkyl-C signal at day 28 for the glucose-amended soil compared with the unamended soil (Fig. 1).

For the alanine-amended soil, the separate methyl-C signal (at 19 ppm) disappeared, but the overall distribution of NMR signal between different shift regions after 28 days was similar to that in the unamended soil. This suggests that the remaining ^{13}C from alanine was distributed between a range of different functionalities.

Table 3. Distribution of NMR signal between different resonances for unamended soil at the outset of the experiment and the soils amended with, ^{13}C$_6$-D-glucose and ^{13}C$_1$-1-L-alanine.

Time and treatment	% of spectrum represented by shift ranges (ppm)				
	10-45	45-90	90-110	110-160	160-200
Unamended soil	32	34	8	16	10
Glucose-amended soil					
0 days	27	43	8	13	8
1 day	25	47	12	15	11
5 days	24	35	9	13	9
28 days	25	32	8	12	8
Alanine-amended soil					
0 days	34	31	9	16	10
1 day	30	32	8	19	10
5 days	29	33	9	19	11
28 days	28	36	9	18	10

4. Discussion

These experiments indicate that ^{13}C from different substrates persist in the soil in different compounds and are indicative of different patterns of substrate utilization. The compounds in which the ^{13}C persists after initial utilization can be regarded as the first generation of intermediates which will subsequently be used as substrates. The fact that different patterns of ^{13}C persistence occur with different initial substrates is an important consideration during protracted substrate utilization assays.

Utilization of glucose leads to relative accumulation of O-alkyl- and alkyl-rich compounds. Consistent observations were made by Baldock et al. (1989, 1990). We cannot assign chemical identities to the residual O-alkyl- and alkyl-C, but structures such as cell walls and cell membranes would be expected to contain such functionalities. The fact that ^{13}C from glucose appears to accumulate in compounds that may be present in structural components of cells is consistent with the larger amount of ^{13}C remaining in the soil (Table 2). By contrast, during the utilization of alanine a larger fraction of the added ^{13}C was lost and the residual ^{13}C was apparently not associated with a particular type of C.

Fig. 1. Intensity-scaled cross polarization magic angle spinning ^{13}C NMR spectra for unamended (a) soil and soil amended with ^{13}C$_6$-D-glucose after 0 (b) and 28 days (c), soil amended with ^{13}C$_1$-1-L-alanine after 0 (d) and 28 (e) days. Numbered peaks correspond to the following classes of C (see also Table 1): carbonyl-C (1), aromatic-C (2), acetal- and ketal-C (3), O-alkyl-C (4), alkyl-C (5), methyl-C (6) and anomeric-C (7).

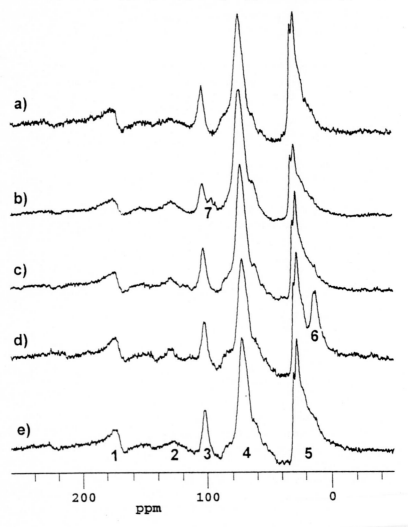

In this work we have demonstrated that the transformations of simple substrates can be followed in soil using NMR. The value of this approach in characterising soil microbial communities on the basis of the substrate utilization and elaboration of metabolites may become clearer when a wider range of soils are contrasted.

5. Acknowledgements

This work was supported by the UK Natural Environmental Research Council.

6. References

Baldock JA, Currie GJ, Oades JM (1991) Organic matter as seen by solid-state 13C nuclear magnetic resonance spectroscopy and pyrolysis tandem mass spectroscopy. In: Wilson WS (ed) Advances in Soil Organic Matter Research. Royal Society of Chemistry, Cambridge, pp 45-60

Baldock JA, Oades JM, Vassallo AM, Wilson MA (1989) Incorporation of Uniformly labelled ^{13}C-glucose carbon into the organic fraction of a soil. Carbon balance and CP/MAS 13C measurements Aust J Soil Res 27:725-746

Baldock JA, Oades JM, Vassallo AM, Wilson MA (1990) Solid-state CP/MAS ^{13}C NMR analysis of bacterial and fungal cultures isolated from a soil incubated with glucose. Aust J Soil Res 28:213-225

Benzing-Purdie L, Cheshire MV, Williams BL, Ratcliffe CI, Ripmeester JA, Goodman BA (1992) Interactions between peat and sodium acetate, ammonium sulphate, urea or wheat straw during incubation studied by ^{13}C and ^{15}N NMR spectroscopy. J Soil Sci 43: 113-115

Hopkins DW, Chudek JA (1997) Solid-state NMR investigation of organic transformations during decomposition of plant material in soil. In: Cadisch G, Giller KE (eds) Driven by Nature: Plant Litter Quality and Decomposition CAB International, Wallingford, Oxon, UK, pp 85-94

Isabella BL, Hopkins DW (1994) Nitrogen transformations in a peaty soil improved for pastoral agriculture. Soil Use Manag 10: 107-111

Kinchesh P, Powlson DS, Randall EW (1995) ^{13}C NMR studies of organic matter in whole soils: I. Quantitation possibilities. Eur J Soil Sci 46: 125-137

Wilson MA (1987) NMR Techniques and Applications in Geochemistry and Soil Chemistry. Pergamon Press, Oxford, UK

Ammonification of Amino Acids in Field, Grassland and Forest soils

Oliver Dilly

Ökologiezentrum, Universität Kiel, Schauenburgerstraße 112, 24118 Kiel, Germany

Abstract. Ammonification rates of different amino acids were studied in order to evaluate the physiology and the function of microbial community in nitrogen cycling in field, grassland and forest soils. Special attention was paid to changes in NO_3^- content during the experimental course and effects of an addition of glucose, NH_4^+ and NO_3^-. The rates of ammonification varied between amino acid used and soil. Highest ammonification rates were observed in case of glutamine addition. However, changes in NO_3^- content were recognised for several soils particularly during the ammonification experiment with glutamine. Addition of glucose and NH_4^+ generally decreased ammonification. In contrast, NO_3^- addition may also increase ammonification rates of amino acids in soil.

Key words. Amino acids, ammonification, arginine, alanine, asparagine, glutamine, nitrate content, glucose addition, microbial physiology, soil

1. Introduction

Amino acids are the largest identifiable group of N containing compounds found in soil (e.g. Hopkins and Ferguson 1994; Burket and Dick 1997). They are readily used as C source by most microorganisms and as N source by those organisms that decompose them to produce ammonia (Richards 1987). In nature, most amino acids are decomposed during the first part of the initial decomposition phase (Marstorp 1996). This underlines the importance of these compounds as substrate for the microbiota.

The ammonification rate of an amino acid has been suggested to be an indicator of microbial activity in soil (Alef and Kleiner 1987; Howard and Howard 1990). However, some of the NH_4^+ formed in the breakdown of amino acids is consumed in the biosynthesis of N compounds (Stryer 1995). Because numerous fungi and bacteria are involved in the breakdown of proteins (Schlegel 1985) and amino acids (Alef and Kleiner 1987) the ammonification rate may be related to the biomass content of the microbiota in soil. No significant differences in biomass-specific ammonification rates of arginine could be detected in agricultural soils that were regularly fertilised with inorganic and organic N sources (Dilly *et al.* 1997). This observation lead to the hypothesis that ammonification rates may be dependent on N status in soils (Dilly and Munch 1995b). Therefore, ammonification rates of different amino acids were studied to evaluate the physiology and the function of microbial community in nitrogen cycling in field, grassland and forest soils. Special attention was paid to changes in NO_3^- concentration during the course of experiment and to the effect of glucose, NH_4^+ and NO_3^- addition on ammonification rates. L-isomers of amino acids were selected because a greater proportion of the microbial community may be capable of using in comparison to D-isomers (Hopkins and Ferguson 1994). The soils under investigations were located in a landscape unit representative for Northern Germany and were part of the "Ecosystem Research in the Bornhöved Lake District" an interdisciplinary approach begun in 1988 to examine and model structures, dynamics and functions of field, grassland, forest and lake ecosystems.

2. Material and Methods

Description of experimental site and soils
The research site is located about 30 km south of Kiel in Schleswig-Holstein, Northern Germany (59°97'N, 35°81'E; Fig. 1). The region ("Ostholsteinisches Hügelland") was formed during the Pleistocene. The climate is influenced by the North Sea and the Baltic Sea. Long-term (1951 to 1980) annual rainfall is 697 mm and average annual air temperature is 8.1 °C.

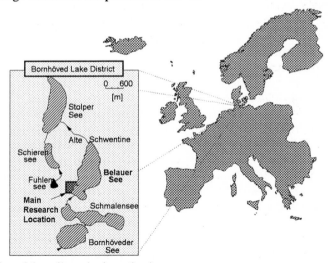

Fig. 1. Location of the soils under investigation

The main terrestrial research area is divided into two catenas, one under agricultural use and one under forest. The soils (Table 1) were predominantly sandy, may have a high organic matter content and varied in pH values. The agricultural soils included: a field with crop rotation (field-CR), regularly fertilised with organic manure and mineral nitrogen; a field with maize (*Zea mays* L.) monoculture (field-MM), regularly fertilised with slurry and mineral nitrogen; a dry grassland regularly fertilised with mineral nitrogen; a wet grassland without fertilisation. The forest soils included: a beech (*Fagus sylvatica* L.) forest and two sites in a black alder (*Alnus glutinosa* (L.) Gaertn.) forest. The pH value of the topsoils under black alder (horizon below litter) differed substantially (Table 1) and the sites located inside the forest at the foot of a hill and directly adjacent to the Lake Belau were therefore called the dystric and the eutric alder forest. The wet grassland and the alder forest sites are located at the shore of Lake Belau. They are seasonally poorly drained and waterlogged.

Table 1. Soil properties of the Bornhöved Lake district, Northern Germany.

	Horizon	Depth	pH	C_{org}	C/N	d_B
		[cm]	[CaCl₂][mg g⁻¹d.m.]			[g cm⁻³]
Agricultural soils						
Field-CR (crop rotation)	Ap	0 - 20	5.5	15.0	9	1.3
Field-MM (maize-monoculture)	Ap	0 - 20	4.4	11.9	10	1.3
Dry grassland	MAh	0 - 10	5.5	18.0	10	1.3
Wet grassland	Aa	0 - 20	5.3	90.0	11	0.6
Forest soils						
Beech	Of	3 - 0	3.4	307.7	19	0.2
	Ah	0 - 5	3.2	29.2	15	1.1
Dystric alder	H	0 - 20	3.7	328.4	15	0.2
Eutric alder	H	0 - 20	5.5	262.1	17	0.2

Sampling procedure

Samples were collected in March 1994 and April 1995. At every sampling, multiple cores were mixed together for each site (50 cores in fields and grasslands with Pürckhauer drill, 20 cores in beech forest and 8 cores in alder forest with a 4-cm diameter drill), gently sieved and stored at 4 °C for maximal 4 weeks. The fraction smaller than 2 mm was used for all subsequent analysis. In beech forest the Of horizon was sampled separately at four different locations using a cylinder (d=15 cm). This was done in order to gain enough material for analyses and to avoid contamination with mineral soil from deeper horizons; the fraction smaller than 5 mm was used. Before these particular samplings, all horizons except Of horizon of the beech forest were sampled from January 1992 to October 1993 at 28-day intervals (n = 22) for the estimation of arginine ammonification and microbial biomass content using fumigation-extraction method. Three subsamples were analysed according to the following description.

Analytical Methods

Ammonification of the different amino acids was determined according to Alef and Kleiner (1986) and modified as follows: Six aliquots of 2.0 g field moist soil (placed in centrifuge tubes of 12 ml) per soil sample were preincubated for 15 min at 30 °C. Then 0.5 ml of a 0.2% (w/v) amino acid solution was added dropwise. Three samples were incubated for 3 h at 30 °C and three were stored at -20 °C. After incubation, 8 ml 2 M KCl were added for extraction, shaken for 15 min and centrifuged for 10 min at 3000 rpm. To 1 ml of the supernant, 4 ml 2 M KCl and 0.2 ml of a 6.71 mM nitroprusside sodium dihydrate / 1.062 M sodium salicylate solution and 0.2 ml of a mixture (4 : 1) of 0.68 M trisodium citrate / 0.5 M sodium hydroxide and 22.66 mM dichloroisocyranuric acid sodium salt dihydrate were added subsequently. After 90 min, the absorbance was determined at 690 nm. For preparation of standards 0, 0.25, 0.5, 1, 2.5, 4, 6 and 8 ml of a 100fold diluted 71.4 mM ammoniumchloride solution were made up to 10 ml using 2 M KCl and then aliquots of 1 ml were treated as samples. NO_3^- was measured at 210 nm using Cu reduction method according to Scharpf and Wehrmann (1976). The following amino acids were used: L-arginine (Arg; Sigma A-5131; hydrochlorid), L-glutamine (Gln; Fluka 49420; BioChemica), L-glutaminate (Glu; Merck 291; for biochemical purposes), L-asparagine (Asn; Fluka 11149; BioChemica; MicroSelect; anhydrous), L-asparagate (Asp; Fluka 11189; BioChemica, MicroSelect), L-alanine (Ala; Fluka 05130; BioChemica), L-leucine (Leu; Merck 5360; for biochemical purposes), L-tyrosine (Tyr; Merck 8371; for biochemical purposes), L-tryptophan (Try; Merck 8374; for biochemical purposes), L-methionine (Met; Fluka 64320; BioChemica). The experiment with glucose addition was carried out with a solution containing 0.2% (w/v) arginine and 0.2% (w/v) glucose (equivalent to 0.5 mg g^{-1} soil). For the treatment receiving NH_4^+ and NO_3^-, the C/N-ratio of the amendment was kept at 30 to ensure complete immobilization of applied N (Azam et al. 1993). Fumgation extraction method for the estimation of microbial biomass C was carried out as described by Dilly and Munch (1995a) with soil-extractant-ratios of 1:4 and 1:16 [dry weight / volume] for the topsoils and Of horizon respectively.

The data were expressed on a soil-volume basis. Three subsamples were analysed for all procedures and standard deviations included in the figures. Statistical analyses were performed with the mean of each site using SigmaStat (Jandel Scientific, Erkrath, Germany). Since the assumption of normality test and equal variance was not fulfilled for the data of the 22 samplings, the nonparametric Kruskal-Wallis One Way Analysis of Variance on Ranks (all pairwise multiple comparison procedures; Student-Newman-Keuls Method or Dunn's Method) was applied to identify differences.

3. Results and Discussion

Arginine ammonification is proposed as a simple method for the estimation of microbial activity in soils (Alef and Kleiner 1986). The ammonification rates of arginine differed between the soils under investigation (Fig. 2). Comparing the agricultural sites, ammonification rates increased in the order, field "maize monoculture" (field-MM) < field "crop rotation" (field-CR) < dry grassland < wet grassland. In forest soils, ammonification was highest in Of horizon of beech forest. It is important to consider that all forest soils had a litter horizon, even if only that of the beech forest was analysed in this study. The ammonification rates of glutamine (Fig. 2), a compound of outstanding significance in N metabolism (Stryer 1995), indicate similar microbial activity for the different sites as do arginine. Whereas arginine ammonification was similar in the topsoils of beech forest and both sites of the alder forest, the wet site of the alder forest showed high ammonification rates of glutamine. This sequences were generally observed for two samplings in March 1994 and April 1995.

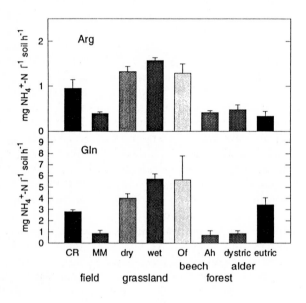

Fig. 2. Arginine (Arg) and glutamine (Gln) ammonification rate in field, grassland and forest soils of the Bornhöved Lake district (for March 1994; n = 3; standard deviations)

A constant ratio between arginine ammonification rates and microbial biomass content that may be referred as qNH_4^+, was observed in topsoils of fields, dry grassland and eutric alder forest (Fig. 3). In contrast, the qNH_4^+ was reduced in topsoil of wet grassland, beech forest and dystric alder forest. This generally indicates that (i) the fraction of the soil microbial community capable of utilizing this amino acid, (ii) the proportion of the amino acid being used for metabolism and (iii) the efficiency in utilizing ammonia differed among the soils.

For the eight soil horizons and the ten amino acids used, significant correlations between ammonification rates were found in most cases (Table 2 and 3). This suggests that most amino acids could be used to indicate the same microbial activity in the soils. Despite this, no significant correlation was observed for alanine (Table 2).

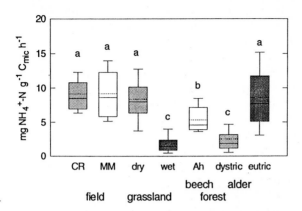

Fig. 3. Arginine ammonification rates related to microbial biomass content in field, grassland and forest soils of the Bornhöved Lake district (Jan 1992 to Oct 1993; n = 22; boxes encompass 25 % and 75 % quartiles, the central and the broken line represent the median and the mean, and bars extend to the 95 % confidence limits; different letters indicate significant differences at p < 0.05)

Table 2. Spearman correlation coefficients between ammonification rates of amino acids in soils of the Bornhöved Lake district (for March 1994; n = 8)

	Ala	Leu	Asn	Gln	Glu
Arg	0.41	0.93**	0.86**	0.90**	0.90**
Ala		0.67	0.60	0.55	0.71*
Leu			0.86*	0.90**	0.98**
Asn				0.86*	0.90**
Gln					0.86*

*, ** Significance of the respective correlation coefficients at p < 0.05, < 0.01

Table 3. Spearman correlation coefficients between ammonification rates of amino acids in soils of the Bornhöved Lake district (for April 1995; n = 8; for Met n = 7)

	Asn	Gln	Met	Tyr	Try
Arg	0.86**	0.88**	0.71	0.90**	0.81**
Asn		0.98**	0.86**	0.86**	0.86**
Gln			0.86**	0.90**	0.88**
Met				0.64	0.86**
Tyr					0.81**

*, ** Significance of the respective correlation coefficients at p < 0.05, < 0.01;

Ammonification rates differed among the amino acids used. The most prevalent amino acid seems to be arginine. The arginine ammonification method was applied to test effects of xenobiotica (Wilke 1989; Hund et al. 1990; Lafrance et al. 1992), to characterise the reclamation or maturity of composts (Insam, 1989; Forster et al. 1993) or to assess the quality and fertility of soils (Franzluebbers et al. 1995). Therefore, ammonification rates of other tested amino acids were related to those of arginine. In comparison to arginine (Table 4 and 5) higher ammonification rates were observed for glutamine and asparagine. Higher rates are even assumed for glutamine when using a higher substrate concentration as used by Frankenberger and Tabatabai (1991). Lower ammonification rates in relation to arginine were observed for alanine, leucine, glutamate, tyrosine, tryptophan, methionine. Methionine interfered in this experimental procedure when determining ammonia content (high background) and, thus, will not be considered. For alanine negative ammonification rates could be observed in wet grassland soil (Table 4). This soil horizon showed a low ratio between arginine ammonification rate and microbial biomass content (Fig. 3) and was therefore considered to be low in N status (Dilly and Munch, 1995b). In addition to the low ammonification-biomass-ratio, a decrease in NO_3^- content could be observed for wet grassland soil during glutamine ammonification (Fig. 4). This indicates that the microbial community may even deplete the inorganic N resources during the ammonification experiment (Fig. 5). Gaseous losses of nitrogen compounds from wet grassland soil seemed to be unlikely because low *in situ* N_2O-emissions were observed at this site (Dilly et al. 1996). With regard to the other achieved data, there was no general pattern of changes in NO_3^- content during ammonification of different amino acids (Table 6). Positive and negative changes in NO_3^- content were observed and should be taken into consideration because the rates may exceed ammonification rates.

Table 4. Ammonification rates of different amino acids related to those of arginine (in percent) in soils of the Bornhöved Lake district (for March 1994)

| | Field | | Grassland | | Forest | | | |
| | CR | MM | Dry | Wet | Beech | | Alder | |
					Of	Ah	Dystric	Eutric
Ala	25	84	39	-22	7	53	38	51
Leu	27	10	18	18	21	22	87	67
Gln	187	126	189	214	49	50	66	349
Glu	47	59	54	49	24	50	80	83
Asp	30	53	24	39	24	49	15	40

The addition of glucose severely reduced ammonification rates (Fig. 6). Consequently, microbial activity present in soil may be underestimated using ammonification rate as an indicator of microbial activity when soluble carbon is present. Furthermore, addition of NH_4^+ and NO_3^- affected ammonification rates (Table 7). Comparing the effect of NH_4^+ and NO_3^-, addition of NO_3^- stimulated ammonification in more cases than NH_4^+ did.

To explore the implication of these experiments for the characterisation of microbial communities in terrestrial ecosystems, soil conditions (i.e. available C and N), the amino acid selected and the product being measured (NH_4^+ and NO_3^-) guided nitrogen liberation rates. High ammonification rates as observed for glutamine and asparagine are desirable since indicating that the substrate is efficiently used. Generally, short-term assays are likely in order to estimate the microbial ecophysiology present at the date of sampling (Brock and Madigan 1991). In order to evaluate the physiology of the microbiota focussing the function in N cycling, both ammonification and change in nitrate content essentially need to syncopate. The sum of NH_4^+ and NO_3^- indicates the net nitrogen mineralisation during the experiment. The different rates of the different amino acids may argue specific biochemical pathways (Stryer 1995) and, therefore, are related to the nitrogen requirement of the microbial communities in soil. From this point of view, the C/N-ratio of each amino acid appears to be of inconsiderable importance. However, more experiments are needed to investigate the distinct varying ammonification rates of the different amino acids.

Table 5. Ammonification rates of different amino acids related to those of arginine (in percent) in soils of the Bornhöved Lake district (for April 1995)

| | Field | | Grassland | | Forest | | | |
| | CR | MM | Dry | Wet | Beech | | Alder | |
					Of	Ah	Dystric	Eutric
Gln	280	212	421	576	321	146	139	224
Asn	228	196	332	442	85	84	102	277
Tyr	26	36	32	80	32	32	7	24
Try	25	75	31	69	50	30	36	46
Met	15	-22	23	88	-201	-76	47	41

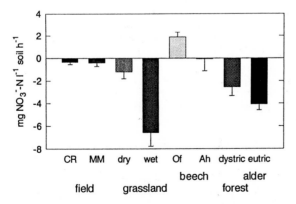

Fig. 4. Change in nitrate content during glutamine ammonification in soils of the Bornhöved Lake district (for March 1994; n = 3; bars represent standard deviations)

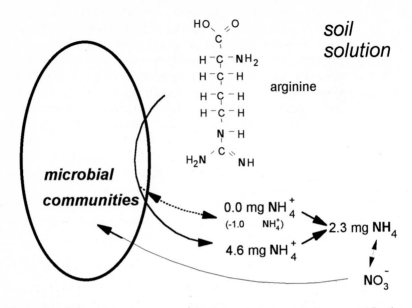

Fig. 5. Fate of inorganic nitrogen compounds in soil during arginine ammonification

Table 6. Change in nitrate content during ammonification experiments with different amino acids in soils of the Bornhöved Lake district (mean of n = 3)

| | Field | | Grassland | | Forest | | | |
| | CR | MM | Dry | Wet | Beech | | Alder | |
					Of	Ah	Dystric	Eutric
Arg [2]	-0.16	0.18	-0.41	-1.54	-0.76	0.15	-0.60	1.36
Leu [1]	0.02	0.17	-0.60	1.06	0.59	0.29	2.55	3.69
Gln [1]	-0.36	-0.39	-1.21	-6.55	1.88	-0.07	-2.54	-4.10
Glu [1]	-0.06	-0.19	-0.51	1.18	0.73	0.18	3.14	3.60
Asn [2]	-0.06	0.73	0.10	-0.83	-2.00	0.29	0.21	-1.75
Asp [1]	-0.43	-0.29	-0.50	2.92	-0.33	0.05	3.59	0.39

[1], [2] for March 1994 and April 1995 respectively

Table 7. Arginine ammonification rates (percent of control) after the addition of NH_4^+ and NO_3^- in soils of the Bornhöved Lake district

| | Field | | Grassland | | Forest | | | |
| | CR | MM | Dry | Wet | Beech | | Alder | |
					Of	Ah	Dystric	Eutric
+ NH_4^+	64	65	82	51	37	43	64	113
+ NO_3^-	135	114	101	67	32	56	58	96

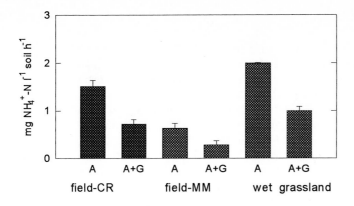

Fig. 6. Addition of glucose (G) on arginine (A) ammonification in soils of the Bornhöved Lake district (for March 1994; n = 3; standard deviations)

4. Conclusions

Comparing the ammonification of ten amino acids in soils of a landscape, the highest rates were observed for glutamine and asparagine. Correlations between ammonification rates of different amino acids and changes in NO_3^- content during experiments were observed. The addition of glucose and NH_4^+ reduced ammonification rates, whereas the addition of NO_3^- also stimulated ammonification in some soils. Ammonification rates of any amino acid may reflect microbial activity. However, rates seem to be controlled by the structure and the physiology of the microbial communities and, thus, by abiotic properties (e. g. N status) in soil.

5. Acknowledgement

This paper is dedicated to Mr. Mirsad Haskovic, who lived as a Bosnian refugee in Kiel, Germany. He carried out most of the experiments presented here. He could not continue his studies because of his schizophrenia that was started with the burning of his asylum house. His struggle ended with his suicide.

6. References

Alef K, Kleiner D (1986) Arginine ammonification, a simple method to estimate microbial activity potentials in soils. Soil Biol Biochem 18:233-235

Alef K, Kleiner D (1987) Applicability of arginine ammonification as indicator of microbial activity in different soils. Biol Fertil Soils 5:148-151

Azam F, Simmons FW, Mulvaney RL (1993) Immobilization of ammonium and nitrate and their interaction with native N in three Illinois Mollisols. Biol Fertil Soils 15:50-54

Brock TD, Madigan MT (1991) Biology of microorganisms. Prentice-Hall Inc. New Jersey

Burket JZ, Dick RP (1997) Long-term vegetation management in relation to accumulation and mineralization of nitrogen in soils. In: Cadisch G, Giller KE (eds) Driven by nature. Plant litter quality and decomposition. CAB International, Wallingford. pp 283-296

Dilly O, Munch J-C (1995a) Microbial biomass and activities in partly hydromorphic agricultural and forest soils in the Bornhöved Lake region of Northern Germany. Biol Fertil Soils 19:343-347

Dilly O, Munch J-C (1995b) Die Argininammonifikation: Bestimmung der mikrobiellen Aktivität oder des Stickstoff-Status in Böden? Mitt Dtsch Bodenkd Ges 75:67-70

Dilly O, Mogge B, Kutsch WL, Kappen K, Munch J-C (1997) Aspects of carbon and nitrogen cycling in soils of the Bornhöved Lake District. I. Microbial characteristics and in situ emissions of carbon dioxide and nitrous oxide of arable and grassland soils. Biogeochem (in press)

Forster J C, Zech W, Würdinger E (1993) Comparison of chemical and microbiological methods for the characterization of the maturity of composts from contrasting sources. Biol Fertil Soils 16:93-99.

Frankenberger WT, Tabatabai MA (1991) L-glutaminase activity of soils. Soil Biol Biochem 23:869-874

Hopkins DW, Ferguson KE (1994) Substrate induced respiration in soil amended with different amino acid isomers. Appl Soil Ecol 1:75-81

Howard PJA, Howard DM (1990) Ammonification as an indicator of microbial activity in soils: Effects of aquaeous tree leaf extracts. Soil Biol Biochem 22:281-282

Hund K, Fabig W, Zelles L (1990) Comparison of methods for the determination of total microbial activity in soil. Agribiol Res 43:131-138

Insam H (1989) Improvement of fly-ash reclamation with organic, organo-mineral and mineral amendments. Proc of the conference `Reclamation, a global perspective`, Calgary, Alberta Vol 2:517-522

Lafrance P, Salvano E, Villeneuve JP (1992) Effect of the herbicide atrazine on soil respiration and the ammonificatin of organic nitrogen in an incubated agricultural soil. Can J Soil Sci 72:1-12

Marstorp H (1996) Influence of soluble carbohydrates, free amino acids, and protein content on the decomposition of Lolium multiflorum shoots. Biol Fertil Soils 21:257-263

Richards BN (1987) The microbiology of terrestrial ecosystems. Longman Scientific and Technical, Essex

Scharpf HC, Wehrmann J (1976) Die Bedeutung des Mineralstickstoffvorrats des Bodens zu Vegetationsbeginn für die Bemessung des N-Düngung zu Winterweizen. Landwirtschaftl Forsch 32:100-114

Schlegel HG (1985) Allgemeine Mikrobiologie. Thieme, Stuttgart

Stryer L (1995) Biochemistry. 4th Edition. Freeman and Company, New York

Wilke BM (1989) Long-term effects of different inorganic pollutants on nitrogen transformations in a sandy Cambisol. Biol Fertil Soils 7:254-258

A New Set of Substrates Proposed for Community Characterization in Environmental Samples

Heribert Insam

University of Innsbruck, Institute of Microbiology, Technikerstr. 25, A-6020 Innsbruck, Austria. *email:* Heribert.Insam@uibk.ac.at

Keywords. Substrate utilisation, Biolog, environment, ecosystem, microbial community.

The selection of substrates offered on commercially available multisubstrate plates (Biolog®, PhenePlate®) is not targeted at environmental samples. In a post-conference workshop of the SUBMECO conference at Innsbruck, Austria (Oct. 16-18, 1996) the need for a set of substrates meeting the requirements of environmental analyses was recognized. Further, a reduction in the number of substrates was advised to allow statistics at a reasonable number of replications (Hitzl *et al.*, 1997b). Also Haack *et al.* (1995) argued that far less substrates than the 95 offered on the Biolog plates are sufficient to distinguish between microbial communities of terrestrial and aquatic ecosystems. According to the information given by Hitzl *et al.* (1997a), Campbell *et al.* (1997), Bochner (pers. comm.) and to previous experience with phenolic compounds (Burkhardt *et al.*, 1993) I proposed a 96-well microplate with 31 substrates plus control, each in 3 replications (Table 1). Such 'EcoPlates' are now available from Biolog®.

Table 1. Substrates offered on the Biolog® EcoPlate (3 replicates of each)

Amines	Carbohydrates	Carboxylic acids	Polymers
putrescine	α-D-lactose	α-keto glutaric acid	α-cyclodextrin
phenyl ethylamine	β-metyl D-glucoside	D-galacturonic acid	glycogen
	cellobiose	D-glucosaminic acid	tween 40
Amino acids	D-mannitol	D-malic acid	tween 80
arginine	I-erythritol	itaconic acid	
L-asparagine	glucose-1- phosphate	methyl pyruvate	**Phenolic comp.**
L-phenylalanine	xylose	γ-hydroxybutyric acid	2-hydroxybenzoate
L-serine	D-galactonic acid lactone		4-hydroxybenzoate
L-threonine	N-acetyl-D-glucosamine		
glycyl-L-glutamic acid	D,L-α-glycerol phosphate		

References

Burkhardt C, Insam H, Hutchinson TC, Reber HH (1993) Impact of heavy metals on the degradative capabilities of soil microbial communities. Biol Fertil Soils 16, 154-156

Campbell CD, Grayston SJ, Hirst D (1997) Use of rhizosphere C sources in sole C source tests to discriminate soil microbial communities. J Microbiol Methods (in press)

Haack SK, Garchow H, Klug MJ, Forney LJ (1995) Analysis of factors affecting the accuracy, reproducibility, and interpretation of microbial community carbon source utilization patterns. Appl Envir Microbiol 61, 1458-1468

Hitzl W, Rangger A, Sharma S, Insam H (1997a) Separation power of the 95 substrates of the Biolog GN plates determined in various soils. FEMS Microb Ecol (in press)

Hitzl W, Henrich M, Kessel M, Insam H (1997b) Application of multivariate analysis of variance and related techniques in soil studies with substrate utilization tests. J Microbiol Methods (in press)

Index

Druck: Strauss Offsetdruck, Mörlenbach
Verarbeitung: Schäffer, Grünstadt

DATE DUE

JUN - 3 1998	
AUG 25 1998	
June 27, 1999	
3/02/00	
JUL 2 3 2001	
SEP 2 2 2006	